ICE-Vol. 8

HISTORY OF THE INTERNAL COMBUSTION ENGINE

presented at

THE ELEVENTH ANNUAL FALL TECHNICAL CONFERENCE
OF THE ASME INTERNAL COMBUSTION ENGINE DIVISION
DEARBORN, MICHIGAN
OCTOBER 15-18, 1989

sponsored by

THE INTERNAL COMBUSTION ENGINE DIVISION, ASME

edited by

E. F. C. SOMERSCALES
RENSSELAER POLYTECHNIC INSTITUTE
TROY, NEW YORK

A. A. ZAGOTTA
SEALED POWER CORPORATION
MUSKEGON, MICHIGAN

THE AMERICAN SOCIETY OF MECHANICAL ENGINEERS
United Engineering Center 345 East 47th Street New York, N.Y. 10017

Statement from By-Laws: The Society shall not be responsible for statements or opinions advanced in papers . . . or printed in its publications (7.1.3)

ISBN No. 0-7918-0375-9

Library of Congress
Catalog Number 89-46164

Printed in U.S.A.

PREFACE

The study of engineering history by the practioners of engineering is not well-developed. This is unfortunate, because, if nothing else, it is the "culture" of our profession, but even more importantly, it provides us with a proper understanding of current and future engineering. Without an adedquate historical background the engineer could, for example, respond incorrectly to problems that might arise in some device or make inappropriate changes in its design. History can also suggest the path that might be followed by a new product, and thereby guide its development and marketing. Because of the fuller appreciation of the art and science of engineering that is provided by an awareness of engineering history, it seems appropriate for the ASME to recognize its role in our profession.

The papers in this volume, which deal with various aspects of the history of the internal combustion engine, were presented in a session at the Fall Technical Conference of the ASME Internal Combustion Engine Division held in Dearborn, Michigan on October 17, 1989. The session was jointly sponsored and arranged by the Internal Combustion Engine Division and by the History and Heritage Committee of ASME. It is the first in what the latter hopes will be a regular series of sessions at various Society meetings jointly sponsored with the different divisions of the Society. It is hoped in this way to raise the consciousness of the engineering community to its history and to encourage in particular the preparation of historical papers by engineer-historians, who are involved in the practice of engineering.

An approximate chronological order has been chosen for the arrangement of the papers, with the first, by H. O. Hardenberg, being on the gunpowder engines, which were experimented with from the sixteenth century to the middle of the nineteenth century. Gunpowder is a difficult fuel to handle and the first practical internal combustion engines, which came into use in the second half of the nineteenth century, were gas engines. Michael F. Marsh's paper deals with their use in the petroleum industry at the turn of the century. Marine diesel engine developments in the early years of this century, where the engines were of substantial power output (up to 2000 BHP per cylinder) are discussed by C. Lyle Cummins, Jr. Toward the end of his paper Cummins describes the remarkable, very large engines that were built for the German navy during the First World War; these were destroyed by the British military authorities in 1919. Charles A. Amann deals in his paper with the history of the spark-ignition engine used in automobiles from the earliest mass-produced engines to current examples. The final two papers by James R. Mondt brings the story up to the present day by reviewing history of emission control methods used in the automotive spark-ignition engine. Although the organizers of the session did not attempt a comprehensive coverage of internal combustion engine history, the topics covered by the papers should give an excellent overview of its historical development.

We would like to acknowledge the part played by the authors of the papers in this book for contributing toward the success of the session. Without their outstanding efforts this publication, which we hope will be of interest and use to engineers, engineer-historians and historians of technology would not have come about.

Euan F. C. Somerscales
A. A. Zagotta

CONTENTS

AN HISTORICAL OVERVIEW OF GUNPOWDER ENGINE DEVELOPMENT – 1508 – 1868

H. O. Hardenberg
Advanced Heavy-Duty Engine Research and Development
Mercedes-Benz AG
Stuttgart, Federal Republic of Germany

ABSTRACT

The internal combustion engine as we use it today originated during the last century, but what is now widely overlooked is that its oldest ancestry dates back even to the Renaissance period.

During the early centuries of this development history, all engines were designed to operate on gunpowder, and it is interesting to note that still during the first half of the 19th century, many efforts were aimed at the application of this type of fuel. Since a complete report on all gunpowder engines would fill a book, the present paper had to be restricted to a concise and exemplary overview of the technical features pertaining to the most important of the manifold development efforts carried out between the years 1508 and 1868.

INTRODUCTION

While the last two centuries of internal combustion engine development have been fairly well studied, the earliest efforts with regard to the development of thermal machinery, namely of gunpowder engines, have not been carefully investigated so far. In this respect, publications on the history of the internal combustion engine are rather incomplete and contain numerous traditional errors, which cannot all be corrected in the framework of this paper. Also, information on the biographies of the researchers and on the state of the science and technology of their times, as well as on the contemporary political events and the cultural background cannot be presented to the desirable extent. Therefore, this paper can give an only limited impression of the efforts of *"turning to the use of Men the great strength of the gun powder, which had hitherto scarce been imploy'd but to their destruction"* (Papin, 1688a).

16TH CENTURY

In ***1508***, ***Leonardo da Vinci*** (1452-1519), the outstanding Italian artist and engineer, sketched the "fire engine", so named by the first words of his short note *"For raising a heavy weight by fire"* on "folio 16 verso" of the so-called Manuscript F (Ravaisson-Mollien, 1881-1891). A rack supported a *"strong vessel of one ell diameter"* (0.595 m) *"and ten ells length"* (5.95 m) with a piston having *"a leather as that of a large bellows"* as a sealing. The engine capacity (cylinder volume minus 10% for the piston) amounted to about 1.5 m³. A fire in the cylinder should be extinguished by closing a lid, so that *"the bottom then will rise from below"*, namely that with decreasing temperature of the gases in the cylinder, the atmospheric pressure would press the piston up and raise a load (Fig. 1).

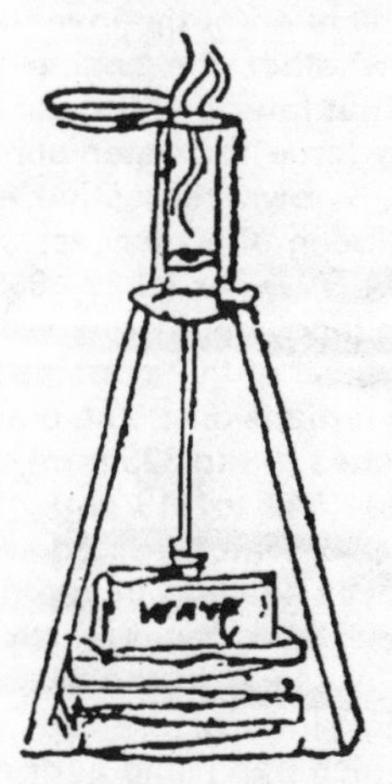

Fig. 1: Leonardo's sketch (true size)

Although Leonardo spoke of *"fire"* only, he may have thought of using gunpowder, as can be concluded from his words *"give a fire from below, as with a bombard"*. With this, the charge was ignited through a narrow bore which appears to have been meant with *"its hole"* - not shown in the sketch - to be closed *"immediately"*. From the rack height being drawn 2.5 times the length of the cylinder, a total height of the engine of about 21 m follows, which compares to that of a seven-storied house. This extraordinarily large machine would have been able to raise a load of 1600 kg to a height of about 3m with the use of about 0.22 kg of powder (Hardenberg, 1988a).

It appears that Leonardo's internal combustion engine activity remained restricted to his sketch and note. Since there is no evidence that the "fire engine" was not an invention of his own, he has to be considered as the inventor of the working cycle of the atmospheric reciprocating internal combustion engine. This statement does not curtail the merits of later engineers who reinvented the process and made use of it in experimental engines - unaware of Leonardo's idea - and it should be noted that the world's first mass-produced internal combustion engine was based on this cycle, namely N.A. Otto's (1832-1891) atmospheric gas engine (Otto and Langen, 1867).

In ***1556***, ***Geronimo Cardano*** (1501-1576), an Italian scientist, proposed the use of gunpowder for propelling road vehicles and boats, but from a merely theoretical point of view, without considering the design of an appropriate engine (Cardano, 1556).

17TH CENTURY

In ***1635***, ***John Babington***, an English mathematician and gunner, and in ***1657***, ***Gaspar Schott*** (1608-1666), a German professor, proposed the application of the force of gunpowder for moving vehicles or raising weights, but neither of them expressed any idea how that might be achieved in practice (Babington, 1635) (Schott, 1657).

In ***1661***, ***Sir Samuel Morland*** (1625-1695), an English diplomat and inventor, petitioned the *"grant for ye sole use"* of his invention of *"an Engin for the raising of water ... in greater quantities, shorter time, & wth much lesse help, then has euer yet been practised or heard of"*. Sir Samuel did not disclose any technical features of his invention, and mention is made of gunpowder neither in his petition (Morland, 1661), nor in the royal warrant for a grant (Charles II., 1661), but a note on the warrant in the State Papers (Green, 1860) speaks of an *"invention for raising water ... by the force of powder and air conjointly"*. No specification of the invention has been found, and it is unknown whether the engine was ever built and tested, but the fact that later on Morland recommended gunpowder as a driving force for water pumps (Morland, 1685) without mentioning his own respective efforts, leads to believe that these had been of no success.

In ***1673***, ***Christian Huygens*** (1629-1695), the Dutch physicist, built several gunpowder engines (Huygens, 1673a to 1673e), all of which were of the same design (Fig. 2), but differed in the cylinder dimensions. The diameters ranged from 2.5 to 12 Parisian inches (68 to 325 mm), the cylinder lengths from 20 to 44 inches (542 to 119 cm). The cylinders were made of tin-plated sheet metal, lined with a thick layer of plaster. The piston was sealed with leather pressed onto the liner surface by means of sponges which were kept wet with water contained in a cavity in the piston. Wetted flexible leather hoses at the sides of the top part of the cylinder served as valves, which hang limp after gunpowder combustion and thus shut the exit ports in the cylinder wall, which were gridded to prevent the hoses from being drawn in. A small vessel charged with gunpowder was screwed into the bottom of the cylinder, right after a piece of tinder with one end reaching into the powder had been kindled.

Evidently, the working principle applied was the same as with Leonardo's engine, while the design differed in several respects, primarily in the direction of the piston motion, which with Huygens' engine required the use of a rope and a pulley to permit the raising of weights. At the time of powder ignition, the piston was *"at the top of the tube"* (Huygens, 1673d) (the usually reported piston upstroke is based on a traditional error). Huygens found that the engine cylinder could be emptied only by about 80%, but did not report of any measured data with regard to the loads and the heights to which they had been raised.

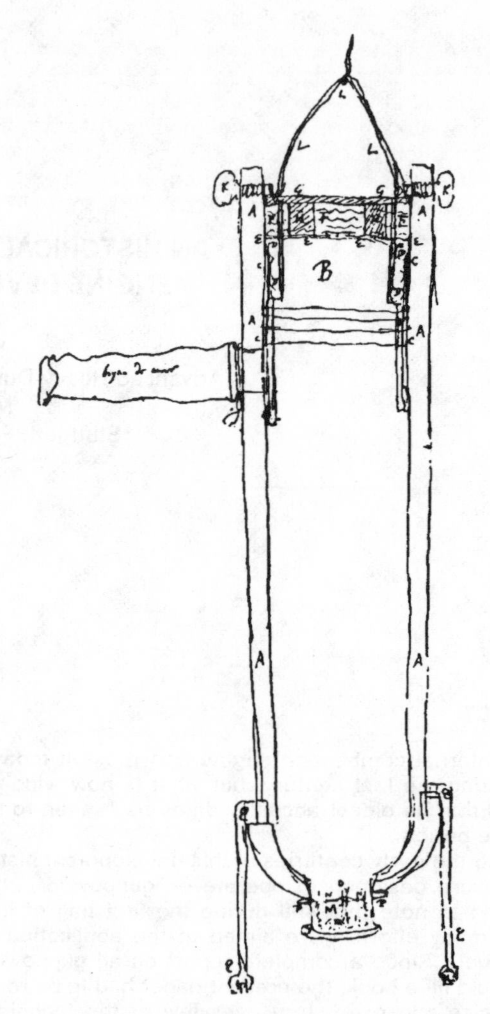

Fig. 2: Huygens' most detailed engine drawing of 1673

In ***1678***, Huygens outlined a piston-less machine for raising water (Huygens, 1678a). In a large space, gunpowder was to be burnt in order to expel the air through a gridded exit port with a leather hose type of valve (Fig. 3). The resulting partial vacuum then would raise water through a 20 to 25 Parisian feet (6.5 to 8.1 m) long one-way valve-equipped tube, which then would flow off through another valve-shut tube - not shown. Perhaps, this note only represents Huygens' reflections on the machine which de Hautefeuille had proposed a few weeks before (Hautefeuille, 1678).

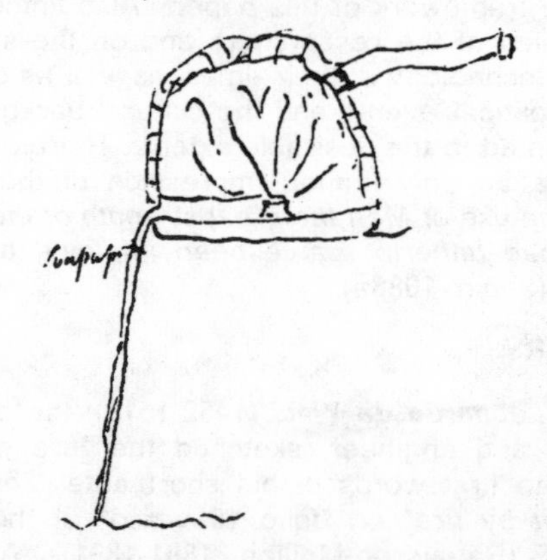

Fig. 3: Huygens' gunpowder water pump of 1678

A few weeks later, Huygens outlined new ideas of an improved reciprocating engine (Fig. 4), but without any text (Huygens, 1678b). Compared to the engines of 1673, the upper sketch exhibits three essential modifications, namely the differing direction of the piston motion, the arrangement of the exhaust valves close to the bottom of the cylinder, and a different position of the powder vessel, which allows a quicker repetition of the engine process in that the recharging can be performed while the piston is still moving. It should be noted that this design is the same as Leonardo's, except for the self-closing valves, the rope and the pulleys.

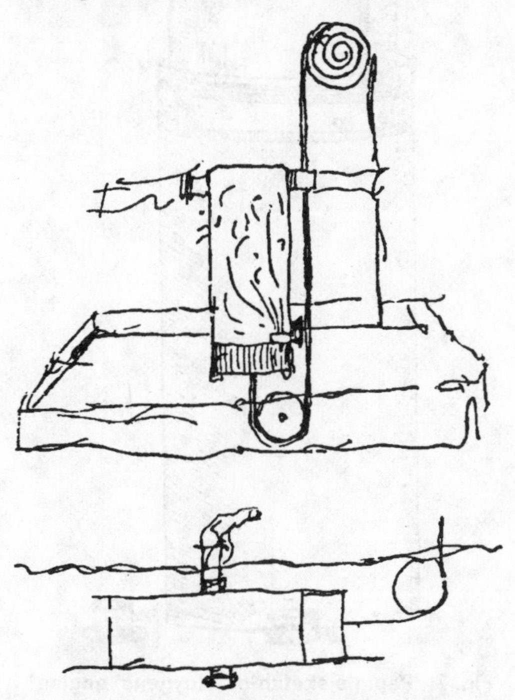

Fig. 4: Huygens' engine ideas of 1678

The lower sketch does not allow definite interpretation. At about half the length of the cylinder, on its upper side, a leather hose serving for a valve can be identified, while the object on the lower side may be interpreted as another - suspending - leather hose or a powder vessel. This arrangement makes sense, if on the unfinished left side, there is a second piston with another rope and pulley, i.e. if it is regarded as a horizontal opposed-piston engine.

It is worth mentioning that after Huygens had stopped his test work by the end of 1673, he did not take it up again, but did remain interested in the gunpowder engine. Evidently, he never came to a final conclusion as to its development potential, since in 1685 he expressed that he *"had not seen enough utility in this invention"* (Huygens, 1685), whereas one year later he wrote that *"this design is not without hope"* (Anon., 1687).

In ***1678***, ***Jean de Hautefeuille*** (1647-1724), a French catholic priest and scientist, proposed the use of gunpowder for raising water using two different methods (Hautefeuille, 1678). According to the first proposal (Fig. 5, left), a vessel *AB* of about 270 to 540 dm³ was to be arranged 30 French royal feet (9.75 m) above the water, into which a tube *CDEF* should reach. By means of a cock *H*, powder should be introduced into the vessel *AB* and was to be ignited after the tap *I* and the valve *D* had been closed. The four automatic valves *G* were intended to allow the exit of the combustion gases and subsequently keep the atmospheric air from entering. After subsequent opening of the valve *D*, water should be raised through the tube *FEDC* into the vessel *AB* as the gases cooled down. Finally, the valve *D* was to be closed and the tap *I* opened for releasing the water from the vessel *AB* into a reservoir - not shown - below the tap *I*. In order to achieve a steadier water flow, it was intended that a second pumping unit consisting of the vessel *KL* with the valve *M* was used alternately. Evidently, de Hautefeuille never built an engine of this type, since if he had performed experiments which he pretended to have made *"on a small scale"*, he would have found that - with a vessel emptied of gas by 80%, as had been obtained by Huygens - the water could be raised only to a height of less than 7.9 m (26 ft).

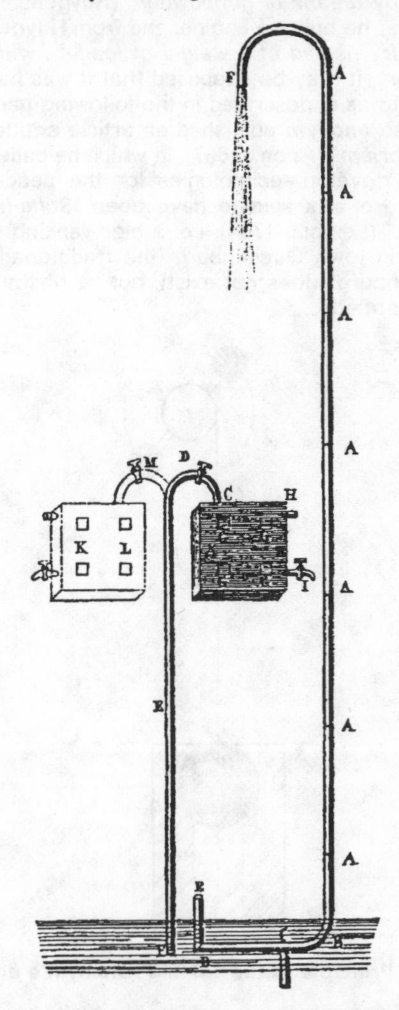

Fig. 5: De Hautefeuille's engine proposals

His second proposal was aimed at raising water to greater heights (Fig. 5, right). In the position *C* below the water surface, a one-way valve should be incorporated in a tube *EDCBAF* of any length, in which at the points *A,A*, one-way valves were to be installed. In the position *E*, there was a cock for the charging with gunpowder. After its ignition, the resulting gas pressure should press the water contained in the tube part *BB* up to any height into a reservoir - not shown - below the tube exit *F*. This proposal was merely theoretical, and it needs no explanation that the system would not have been practical. It is erroneous that in 1682, de Hautefeuille made further gunpowder engine proposals, since in the respective paper (Hautefeuille, 1682) he described a new type of water pump and made mention of gunpowder engines only in passing, primarily of Huygens' engine.

There can hardly be any doubt about the fact that neither of de Hautefeuille's engine proposals has ever been built. Nevertheless, he has to be regarded as the inventor of the piston-less atmospheric gunpowder engine, as long as it is not known whether Morland's engine was of the same type, and since Huygens' respective ideas are of a later date. Also, de Hautefeuille is the inventor of the piston-less direct-acting gunpowder engine.

By ***1683***, ***Wilhelm Heinrich Freyberg*** (1617-1696), a German high-ranking functionary of the principality of Anhalt-Dessau, tried *"to find the invention of grinding grain and other things by means of gunpowder"* (Huygens, 1888-1950). It appears that he built an engine, and from Huygens' remark *"I raise but air, instead of a weight of lead, ... with the effort of the powder"*, it may be supposed that it was based on the same principle as is described in the following paragraph.

In ***1687***, an ***anonym*** published an article entitled *"Ad Majorem Dei Gloriam"* (Anon., 1687), in which he called upon the scientists to develop technologies for the peaceful use of gunpowder. He was said to have been *"Stifts-hauptman à quedlinbourg"* (Leibniz, 1704), i.e. a high-ranking functionary in the German town Quedlinburg (the traditionally reported place "Zödtenburg" does not exist), but he has not yet been definitely identified.

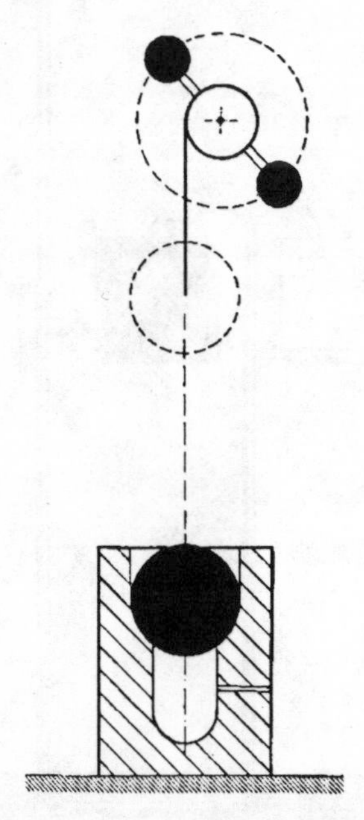

Fig. 6: Principle of the *"AMDG"*-anonym's engine

Of his engine which was obviously built, he wrote that *"half an ounce of powder"* (14.6 g) *"drives a weight of 15 pound"* (7 kg) *"to a height of 16 to 18 feet"* (4.5 to 5.1 m) *"and connects itself to a rope, which is tied to the axle of a small mill wheel"*, which *"with the gravity of the weight makes two small stones of 2 or 3 pound"* (0.93 or 1.4 kg) *"go round 160 times, until the said weight gets, from where it was set in motion and where it expects another impulse"* (Fig. 6). These results correspond to an efficiency of only less than 1%, but it has to be taken into account that the *"AMDG"*-anonym considered his engine only as a *"simple and plainly useless experiment"*, i.e. demonstration model.

In ***1687***, ***Denis Papin*** (1647-1712?), a French physician who had been Huygens' assistant, read a paper to the Royal Society of London (Papin, 1687), in which - refering to Huygens' test results reported by the *"AMDG"*-anonym - he pretended to describe Huygens' engine. Actually, he outlined an engine (Fig. 7) which - apart from the lever for applying the powder vessel - was identical to the design which he published as his own invention a year later (Fig. 8). It is not clear why Papin gave a conspicuously false description of Huygens' engine.

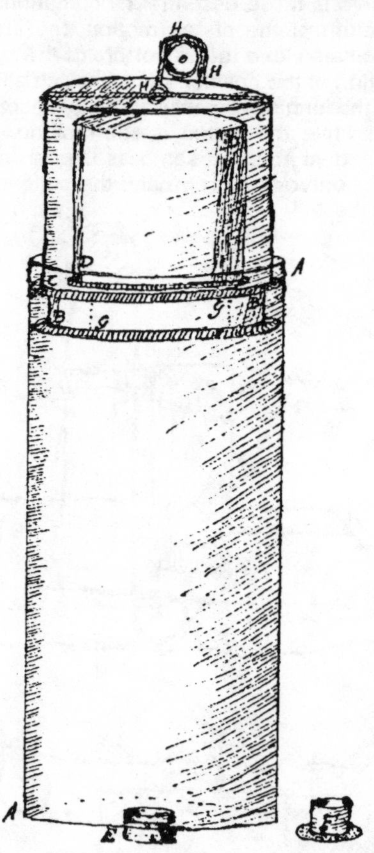

Fig. 7: Papin's sketch of "Huygens' engine"

In ***1688***, after having become Professor of Mathematics at the University of Marburg, a town in the German county of Hesse-Cassel, Papin wrote nearly identical papers on his gunpowder engine in three different languages. The French and Latin versions were published (Papin, 1688b and 1688c), whereas the English text (Papin, 1688a) was only read to the Royal Society. The cylinder diameter was 5 inches and the length 16 inches, which, if local units were used, would compare to 120 and 384 mm, respectively. The traditionally reported dimensions of 130 and 400 mm are incorrect, particularly since they would imply different conversion factors for the diameter (26 mm/in) and the length (25 mm/in). The cylinder was made of *"a hamer'd brass plate"* with a flat bottom (Fig. 8). One of the main design features was the exhaust valve in the piston. It is obvious that with this arrangement, the traditionally reported upstroke of the piston cannot occur, since with the piston in its lower position, the combustion gases would have left the cylinder through the valve without building up a pressure high enough to move the piston upwards. Also, the charging was simplified, in that the powder vessel was sealed with a water-wetted leather and pressed against the cylinder bottom by means of a weight-loaded lever.

The engine, which he considered to be superior to Huygens', had been improved in the design, but not with regard to effectiveness. Having expected to achieve a complete vacuum in the cylinder - as had Huygens who had succeeded in emptying his engine only to 1/6 - he found *"that ther' remains in it about 1/5 of the air that it do'th usually contain"*. One reason for this poorer result may have been that at the end of the powder combustion in Papin's cylinder, there remained a pressure slightly higher than atmospheric, because of the weight of the exhaust valve and thereby caused flow resistance, which prevented a more complete vacuum. The figures given of the weight of *"150 pounds"* (70 kg) which the engine could raise to a height of *"1 foot"* (29 cm) were not measured, but merely *"esteem'd"*. Obviously, this was done incorrectly, because this result would have required emptying the cylinder to the unachievable extent of 93%. Actually, the engine would have been able to raise the weight of 150 pound to a height of only about 1/2 foot, which still corresponds to a quite considerable efficiency of 4%, but only to a rather poor brake mean effective pressure of 0.2 bar (3 psi) (Hardenberg, 1988b).

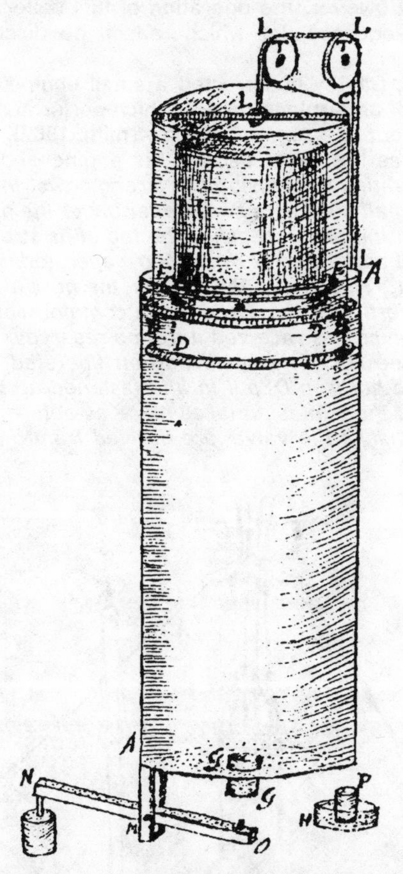

Fig. 8: Papin's sketch of his own engine

Papin finished his experimental work, of which he had done *"but very little"*, after about half a year; in fact, it is doubtful if it had been performed at all, since there is some evidence which suggests that it had not. At that time, he may have begun to realize the power potential of condensing or expanding steam, which he made use of in his atmospheric steam engine of 1690 and the high-pressure version of the years between 1698 und 1707, respectively.

18TH CENTURY

Possibly due to the tremendous progress of steam engine technology during the 18th century and the resulting overestimation of its development potential, a break of more than hundred years in the development of the gunpowder engine occured, after Papin had abandoned his respective efforts.

In ***1753***, ***Daniel Bernoulli*** (1700-1782), a Swiss Professor of Physics, discussed the naval application of gunpowder engines from a merely theoretical point of view, but did not elaborate on their design (Bernoulli, 1753).

That in ***1783***, ***Laurent-Gaspard Gérard***, a French flight enthusiast, built a gunpowder engine is incorrect. Actually, he only discussed its application in airplanes, hoping that a *"man of genius"* might invent it (Gérard, 1784).

19TH CENTURY

In ***1800***, ***George Medhurst*** (1759-1827), an English engineer, was granted a patent (Medhurst, 1800) for compressed air-powered vehicles, in which a gunpowder engine was also described. With this direct-acting expansion machine, having a cylinder capacity of 26.6 cubic inches (437 cm^3) (Fig. 9), the connecting rod acted through a parallel-motion mechanism to a crankshaft - not shown. For a power output of one horsepower, 1/15 cubic inch (1.09 cm^3) of powder should be introduced from a funnel by means of a stopcock, as the piston approached its top dead center (TDC). This stopcock was to be rotated by the crankshaft with half its speed by means of a gearing. The ignition should be effected by means of a spark-producing flint wheel type of mechanism, actuated by the piston when reaching TDC - not shown, since only basically described. The air enclosed between the piston and the bottom of the cylinder, which was equipped with a stuffing-box, should *"act as a spring to return the piston"*. A small spring-loaded inwardly opening valve in the bottom was expected, erroneously, to be opened by the atmospheric pressure at TDC position of the piston. With the piston in its bottom dead center position, the pressure in the working cylinder should be lower than atmospheric, which would cause the valve in the cylinder head to be opened and permit the combustion gases to be exhausted until the piston would shut the valve when approaching TDC.

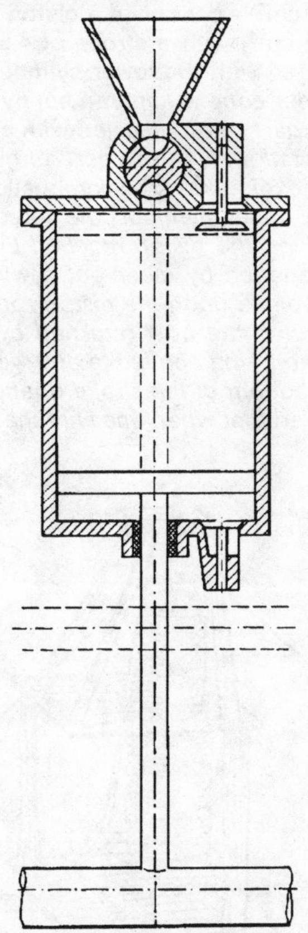

Fig. 9: Medhurst's engine cylinder

Although this is not a practical design and was apparently never built, it nevertheless represents the oldest draft of an internal combustion reciprocating piston crank engine with positively controlled cyclic repetition of the working process.

In ***1806***, the French brothers ***Joseph Claude*** (1763-1828) and ***Joseph Nicéphore Niépce*** (1765-1833) were granted a patent (Niépce, 1806) on a strange, turbine-like, direct-acting internal combustion engine for which they named gunpowder as a potential fuel, amongst nearly all other liquid and solid combustibles. However, the engine was not designed for the use of gunpowder, and this fuel does not appear to have been closely considered let alone applied.

In ***1810***, a Frenchman named ***Henri*** proposed a gunpowder-operated ram (Anon., 1810) which here is only mentioned for the sake of completeness. Not a gunpowder engine in the usual sense, it was a cannon-type of direct-acting machine of impractical design - modern rams work indirectly in that the combustion of fuel raises a weight which falling down effects the ramming.

In ***1807***, ***Isaac de Rivaz*** (1752-1829), a Swiss inventor, obtained a patent for a machine (Rivaz, 1807), which was no gunpowder engine in the true sense of the words, in that the engine power was obtained exclusively from powder combustion. Actually, only the cylinder charge - of coal gas and air - was ignited with gunpowder in that *"some grains ... drop onto a bottom kept red-hot"*.

In ***1807***, the English baronet ***George Cayley*** (1773-1857), inventor of the rigid airplane, built a small gunpowder engine *"for the purpose of some preparatory experiments on aerial navigation"* (Fig. 10) (Cayley, 1799-1826). Two cylinders were connected with each other, of which the lower - with a capacity of 117 cubic inches (1917 cm³) - was *"merely a reservoir of air"*, and of which the upper - with a capacity of 40 cubic inches (655 cm³) - contained a piston of a volume of 7.9 cubic inches (129 cm³) with a stroke of 4.9 inches (124 mm) (Fig. 10). Connected with the lower cylinder *"was a tube terminated in a slender cone ... kept red hot by a lamp. A branch above the sloping part communicated with a small cock, which at each stroke let fall a given measure of gunpowder into the hot pipe, where it exploded"*. The combustion gases operated *"the piston in the upper cylinder, the force of which bent a strong bow ... When the piston reached the top of its stroke a stop cock was opened by means of a wire passing thro' a loop in the cross piece upon the piston rod. This let out the elastic vapour & then the bow returned by its elasticity, the power of the stroke being applied to any work proposed. The same wire at the bottom of the stroke opened the gunpowder cock, ... so contrived that when one shut the other opened and vice versa"*.

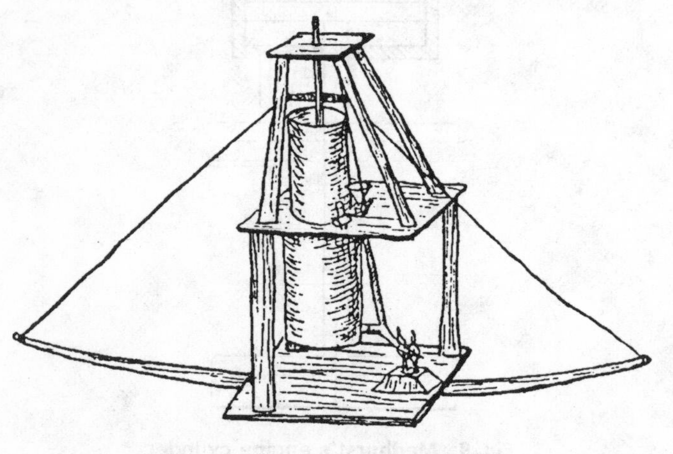

Fig. 10: Cayley's engine of 1807

The force of the bow was 50 pound (22.7 kg), which to be achieved, 5.5 grains (0.36 g) of powder were required. From this result, and for a stroke taking place each second, Cayley calculated a powder consumption of 63 lb/BHP-h (39 kg/kWh) - which corresponds to an engine efficiency of 3.3% - and conceived *"that the reservoir was too large and that with half the size the force would have been much greater"*. Sir George abandoned this idea and did not take up work on the same type of gunpowder engine until 1850.

In ***1820***, Cayley tried a different type of engine, in which he fired *"gunpowder in a long case thro' water so as to generate both air and steam"* (Pritchard, 1961). In other words, powder combustion gas and steam formed in an *"odd sort of boiler"* were expanded in the cylinder of a steam engine-like machine. However, the operation of this boiler appeared to him too dangerous, for which reason he discontinued the work on it.

In ***1850***, Cayley constructed a small engine for vibrating the wings of an airplane model, which performed *"the up and down in about one second"* (Gibbs-Smith, 1962). The general principle was the same as with his engine of 1807, namely *"that of elevating the tension of a strong bowstring A,A, by the blast of a small portion of gunpowder under the piston B"* (Fig. 11). *"When the piston reaches the top of its stroke a valve is opened in it and the power of the bow operated freely causing the descent. When near the bottom, the gunpowder is liberated by inverting a cup made like a common cock for letting off liquor which has received it from a reservoir when turned upwards (see C). The small portion liberated falls upon a piece of red-hot iron D, put in and fastened up by the screw door E, and the explosion takes place as before. This movement and that of the valve, are derived from that of the engine.*

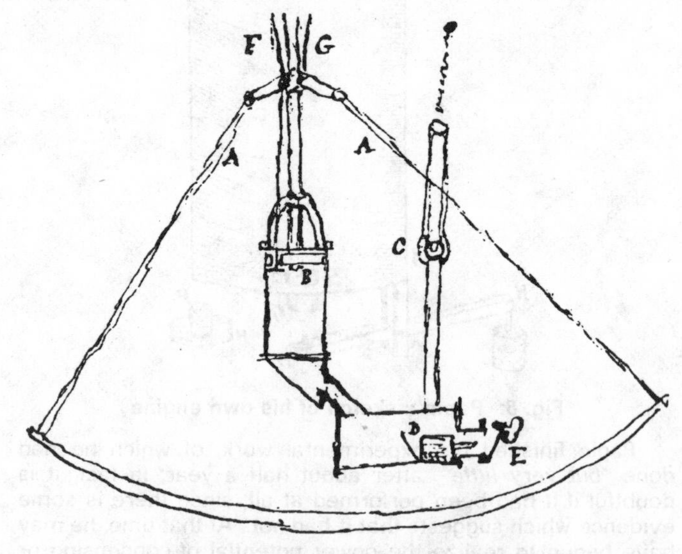

Fig. 11: Cayley's engine of 1850

The test data do not allow thermodynamic evaluation, but Sir George appears to have been satisfied with the results, since at about the same time, he designed a gunpowder-engined monoplane aircraft model, but there is no note of it being tested in flight, if indeed it was ever built.

That in ***1817***, ***William Farish*** (1759-1837), an Anglican priest and Cambridge Professor of Chemistry, constructed a small engine intended to be driven by gunpowder is unlikely and could have been confused with the proposal mentioned in the following paragraph.

In ***1820***, ***William Cecil*** (1792-1882), also an Anglican priest, presented a paper describing a hydrogen-fuelled atmospheric engine (Cecil, 1820). In this, he also mentioned a machine operated by gunpowder, which was a self-operating atmospheric engine, of which no design details were given, and which was never built.

In *1823*, ***Jacques-Philippe Mérigon de Montgéry***, captain of the French Imperial Marine, discussed the propelling of submarine war ships (Montgéry, 1823). Aware of the fact that for the combustion of powder, no air is required, he considered gunpowder engines to be most suitable for this purpose, and after reviewing earlier efforts with regard their development, he proposed an engine of his own design.

A circular magazine, similar to those in automatic weapons, of 30 powder charges of 0.64 g (9.8 gr) was intended to be used in order to achieve sufficiently fast repetition of the engine operation cycle. To ignite the powder, two small hammers should beat caps on both sides of every powder charge. In a cylinder of 1 French inch (27.1 mm) diameter and 3 inches (81.2 mm) length, a piston should be moved up by the powder combustion, directly acting on the boat-moving mechanism. Then, the cylinder should be opened at the bottom, and the piston would fall back for another working cycle. It was intended to have two of the systems described, linked in such a way that when the one piston went up, the other would go down. This idea remained merely theoretical, and an engine of this design was never built.

In *1833*, the American ***Henry Rodgers*** applied for a patent, of which the specification appears to be lost. A piston was *"forced up by the explosion of gunpowder, ignited by means of a hammer striking upon percussion powder"* (Jones, 1833), obviously, very similar as in Montgéry's engine. Actually, he did not claim the machine, but the principle of using *"the power produced by the explosion of gunpowder or any other explosive compound, to machinery in general"*, which was not patentable for lack of novelty.

In *1837*, the Scotsman ***J. Smith*** claimed to have built a *"gunpowder engine, and which moves ... against a weight of twenty-six hundred weight on the square inch of the piston"* (179 bar) and calculated the fuel saving compared with steam *"to be fully eighty per cent, whilst the space it occupies is not one-twentieth of that taken up by the steam-engine"* (Anon., 1837). From the short press note, it can only be concluded that it was a steam engine-like expansion machine, may be similar to Medhurst's or to the one described in the following paragraph. Evidently, the engine did not meet the inventor's expectations, since there was no further news of it.

In *1837*, the Englishman ***William Horatio Potter***, proposed an engine consisting of a combustion chamber *A, "into which is introduced, at regular intervals, by the rotation of a stopper ... with a cavity of convenient form in it, a certain quantity of gunpowder, to be supplied constantly. This will fall into the cup B, at the bottom of which are two small brass knobs at a proper distance, communicating with the wires C,D, which proceed from the positive and negative surfaces of a small Leyden vial, which is kept constantly charged by the revolution of a small electrical plate. Now it is easy to see how both plate and stopper ... may be made regularly to revolve by the engine itself, and that with the expense of scarcely any power. SV, is a safety-valve. E is the pipe to convey the power from the generator A, to the cylinder, which is the same as that used in high pressure engines"*. (Potter, 1837). Compared to Medhurst's engine, Potter's machine was more advanced only with regard to the application of the electrical ignition, whereas the use of a separate combustion chamber represented a step back in development.

In *1868*, ***Robert Charles Jay*** an English engineer, exhibited two engine models, namely a *"Rotary Engine of steel plate, one-horse power, weight 60lbs., motive power gun-cotton"* and a *"Model of Gun-cotton Engine"* (Anon., 1868). It can be assumed that Jay prefered the use of guncotton, i.e. cellulose nitrate, which only contains carbon, hydrogen, oxygen and nitrogen, because of its smoke-less combustion, while the gunpowder components - potassium nitrate (ca. 75%), charcoal (ca. 15%) and sulfur (ca. 10%) - cause quite considerable residues in engines. No design features of the engines are known, nor whether they were actually workable, since *"none of these were shown in motion"*, and it appears that Jay's efforts were the last with regard to the use of explosives in internal combustion engines.

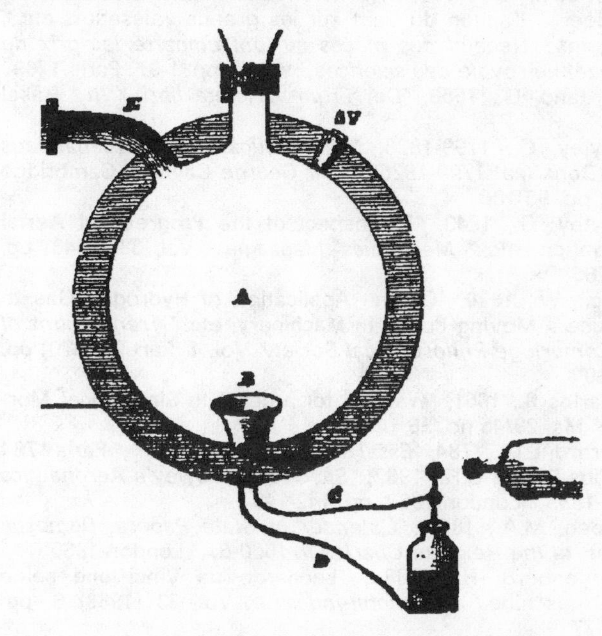

Fig. 12: Potter's gunpowder gas generator

SUMMARY

Apart from Leonardo's activities, there are two main phases of gunpowder engine development.

The first phase, i.e. the 70s and 80s of the 17th century, may be called the pre-steam period, and is characterized by the search for power sources for raising weights. For this purpose, gunpowder seemed to be highly attractive as a fuel, because the violence of its reaction made it appear more powerful than any other combustible available; and also the fact that gunpowder does not require any air for its combustion may have played a role in the unusual interest in its use.

The second phase, i.e. the first half of the 19th century, or the post-steam period, is characterized by the search for engines and fuels suitable for mobile purposes. It was stated fairly early that *"many essays have been made to apply the explosive force of gunpowder to the propelling of machinery, and after the expenditure of large sums of money, it has been abandoned as hopeless"* (Jones, 1833), and also: *"Gunpowder is too dangerous, but would, at considerable expense, effect the purpose; but who would take the double risk of breaking their necks or being blown to atoms?"* (Cayley, 1843). However, efforts were nevertheless continued, particularly by airplane enthusiasts, and it is interesting to note that the last machine of this kind, i.e. Jay's engine, was developed after mass production of N.A. Otto's famous atmospheric gas engine had been started.

REFERENCES

Anon., 1687, "Ad Majorem Dei Gloriam," *Nouvelles de la République des Lettres*, Mai 1687, VI, pp. 516-523

Anon., 1810, "Machine propre à enfoncer les pieux par l'effet de la poudre," *Annales des arts et metiers,* (1810) pp. 311-312

Anon., 1837, "Gunpowder Engine," *Mechanics' Magazine,* Vol. 28 (1837) p. 159

Anon., 1868, "*Report of the First Exhibition of the Aëronautical Society of Great Britain ... 1868, etc,*" Greenwich 1868

Babington, J., 1635, "*Pyrotechnia or, A discourse of artificiall fire-works, etc.,*" London 1635

Bernoulli, D., 1753, "Recherches sur la maniere ... de suppléer à l'action du vent sur les grands vaisseaux, etc.," 1753, in: "*Receuil des pieces qui ont emporté les prix de l'Académie royale des sciences,'* Vol. 7, pp. 1-97, Paris 1769

Cardano, G., 1556, "*De rerum varietate libri XVII,*" Basel 1556

Cayley, G., 1799-1826, "*Aeronautical and Miscellaneous Note-Book (ca. 1799-1826) of Sir George Cayley,*" Cambridge 1933, pp. 95-100

Cayley, G., 1843, "Retrospect of the Progress of Aerial Navigation, etc.," *Mechanics' Magazine* - Vol. 38 (1843) pp. 263-265

Cecil, W., 1820, "On the Application of Hydrogen Gas to Produce a Moving Power in Machinery, etc.," *Transactions of the Cambridge Philosophical Society,* Vol. 1, Part 2 (1820) pp. 217-39

Charles II., 1661, "Warrant for a grant to Sir Samuel Morland," Ms. 29/45 no. 36

Gérard, L.G., 1784, "*Essai sur l'art du vol aérien,*" Paris 1784

Gibbs-Smith, C.H., 1962, "*Sir George Cayley's Aeronautics 1796-1855,*" London 1962, pp. 142-143

Green, M.A., 1860, "*Calender of State Papers, Domestic Series, of the Reign of Charles II. 1660-61,*" London 1860

Hardenberg, H., 1988a, "Leonardo da Vinci und seine Feuermaschine," *Automobil-Industrie,* Vol. 33 (1988) 6, pp. 671-677

Hardenberg, H., 1988b, "Denis Papin und sein Schießpulvermotor," *Automobil-Industrie,* Vol. 33 (1988) 3, pp. 279-285

de Hautefeuille, J., 1678, "*Pendule perpetuelle, avec un nouveau balancier; et la maniere d'élever l'eau par le moyen de la poudre à canon, etc.,*" Paris 1678

de Hautefeuille, J., 1682, "*Reflexions sur quelques machines à élever les eaux, etc.,*" Paris 1682

Huygens, C., 1673a, "Nouvelle force mouvante par le moyen de la poudre a Canon," Ms. D, pp. 327-329, February 10, 1673

Huygens, C., 1673b, "Pour faire les tuyaux," Ms. D, pp. 325-326, March 1673

Huygens, C., 1673c, "Nouvelle force mouvante par le moyen de la poudre a canon et de l'air," Chartæ Mechanicæ, pp. 88-89, August 26, 1673

Huygens, C., 1673d, "Letter to his brother Lodewijk," September 22, 1673

Huygens, C., 1673e, "Test report," Ms. D, p. 330, December 23, 1673

Huygens, C., 1678a, "Raising water by means of gunpowder," Ms. E, p. 133, August 30, 1678

Huygens, C., 1678b, "Gunpowder engine sketch," Ms. E, pp. 144-145, September 11, 1678

Huygens, C., 1685, "Letter to Vegelin van Claerbergen," April 19, 1685

Huygens, C., 1888-1950, "*Œuvres complètes,*" 22 Vols. Den Haag 1888-1950, Letter Nos. 2330, 2382, 2495, 2496, 2498

Jones, T.P., 1833, "For the application to machinery of the Power resulting from Explosive Compounds, etc.," *Journal of the Franklin Institute,* Vol. 7 (1833) 6, pp. 409-410

Leibniz, G.W., 1704, "Letter to Papin," Ms. No. 29909, April 11, 1704

Medhurst, G., 1800, "Obtaining Motive Power, Propelling Carriages, &c.," British Patent 1800 - No. 2431

de Montgéry, J.P.M., 1823, "Mémoire sur la navigation et la guerre sous-marines," *Annales maritimes et coloniales,* (1823) Part 2, Vol. 2, pp. 292-355

Morland, S., 1661, "Petition to King Charles II.," Ms. 44/48 f.1, December 1661

Morland, S., 1685, "*Elevation des eaux ... par le moyen d'un nouveau piston ... d'un mouvement cyclo-elliptique,*" Paris 1685

Niépce, J.C.; Niépce, J.N., 1806, "Pyréolophore," French Patent 1806 - No. 409

Otto, N.A., Langen, E., 1867, "Atmosphärische Gas-Kraftmaschine," *Dinglers Polytechnisches Journal,* Vol. 186 (1867) 10, pp. 90-97

Papin, D., 1687, "Mr. Papin about the using Gunpowder to raise weights," Register of the Royal Society, Vol. VII (1687) pp. 24-25

Papin, D., 1688a, "Doctor Papin about raising weights by the force of Gunpowder," Register of the Royal Society, Vol. VII (1688) pp. 27-32

Papin, D., 1688b, "Mémoire ... touchant un nouvel usage de la poudre à Canon, etc.," *Nouvelles de la République des Lettres,* September 1688, III, pp. 982-991

Papin, D., 1688c, "Excerpta ex ... litteris ad --- de Novo Pulveris Pyrii usu," *Acta Eruditorum,* September 1688, pp. 497-501

Potter, W.H., 1837, "Potter's Gunpowder Engine," *Mechanics' Magazine,* Vol. 28 (1837) p. 146

Pritchard, J.L., 1961, "*Sir George Cayley - The Inventor of the Aeroplane,*" London 1961, p. 109

Ravaisson-Mollien, C., 1881-1891, "*Manuscrits de Léonard de Vinci,*" 6 Vols. Paris 1881-1891

de Rivaz, I., 1807, French Patent 1807 - No. 731

Schott, G., 1657, "*Magia universalis naturæ et artis, etc.,*" Würzburg 1657

EARLY GAS ENGINES IN THE PETROLEUM INDUSTRY AT THE TURN OF THE CENTURY

M. F. Marsh
Cooper Industries
Springfield, Ohio

ABSTRACT

By the late 1800's the increased use of petroleum products increased the use of engines in the oil fields. Steam was king, but a new style of engine was becoming prominent. New engine companies were started, and existing companies were beginning to manufacture internal combustion engines. The newly designed and developed engines would run on well head gas, and there was no need for a boiler. This paper depicts some of the early engine innovations and improvements to the natural gas industry at the turn of the century.

INTRODUCTION

The American Petroleum Industry started in 1859 when Col. Edwin Drake and the Seneca Oil Company drilled for oil in northern Pennsylvania. The well produced a low cost source of oil at a time when demand was starting. The main use of petroleum was as a lubricant. Demand for oil rapidly grew by 1500 times during the first 3 years. With this rapid increase in use of oil came a demand for engines to pump or drill for the oil.

Companies like Ajax Iron Works, C and G Cooper, and Snow Steam Pump Works were there meeting the demands of the new oil companies and filling the roll of supplying steam engines for drilling and pumping. After all, the steam engine was an established form of power.

Early oil production also saw the development of a new engine style: the internal combustion engine. Escaping wellhead gas was considered a waste by-product for the oil producers. Many of these oil wells were pumped with steam engines through the early 1890's. Unfortunately, production was beginning to fall to a point where the cost of pumping with steam became increasingly prohibitive. Old boilers were decaying from hard water and needed to be replaced. Large water supplies were needed for steam while natural gas escaped uselessly into the air.

Several innovative people decided to make the new internal combustion engine run on the wellhead gas by-product. This would give the oil producers a new,less expensive means to pump oil. Companies like Reid, Superior, Bessemer and others started producing internal combustion engines.

The internal combustion engine of the late 1800's can be divided into two major classifications. The first is the 4-cycle developed by Nicolaus Otto. The second is the 2-cycle developed by Sir Dugald Clerk. A sub-classification for both styles is the type of ignition. Three major styles of ignition were used. The most common for engines was the spark ignition. The second style was the hot-tube. The third was an open flame or flame cock ignition. All three styles found their place in the oil industry.

Figure 1
Early Oil Pumping Typical
at the Turn of the Century

Figure 2
Ajax Steam Engine to Power Pump Rig

Otto Gas Engine Works, Philadelphia, PA.

One of the first gas engine manufacturers to make engines for the oil field was the Otto Gas Engine Works. The Engine Works was formed by the Schleicher brothers to sell internal combustion engines under U.S. Patent 194047. The Schleicher brothers were relatives of Eugene Langen of Otto & Langen engine and formed Schleicher, Schumm & Company was formed in 1880. In 1894, the interests of the company were sold to Gasmotors Fabrile Dentz and became known as The Otto Gas Engine Works.

In 1884, the engine works produced a line of single cylinder models from 1 to 25 horsepower. They also manufactured two cylinder models rated up to 100 horsepower. Most of the engines manufactured by Otto were horizontal, however,a vertical style was also made and was offered in sizes from 1/2 to 4 horsepower. By 1912 the company built engines up to 1000 horsepower.

Otto claimed to have been first with the four cycle engine - this was an irrefutable fact. Other "firsts" claimed in the 1894 catalog were: adoption of poppet valves, development of hot tube igniter, first reliable electric igniter, first to use spiral gears in the lay shaft drive, and many other unique features.

The new company prospered until World War I. The U.S. Government expropriated the Philadelphia assets under the "Trading with the Enemy Act." The company was then sold to Superior Gas Engine Company of Springfield, Ohio in 1923 and continued to operate as the Otto Works of Superior Gas Engine Company.

The early Otto engines were the first to use the 4-cycle concept. Along with the 4-cycle design, they used a unique valve and ignition arrangement. The early Otto cycle engines used a flame ignition system and slide valve.

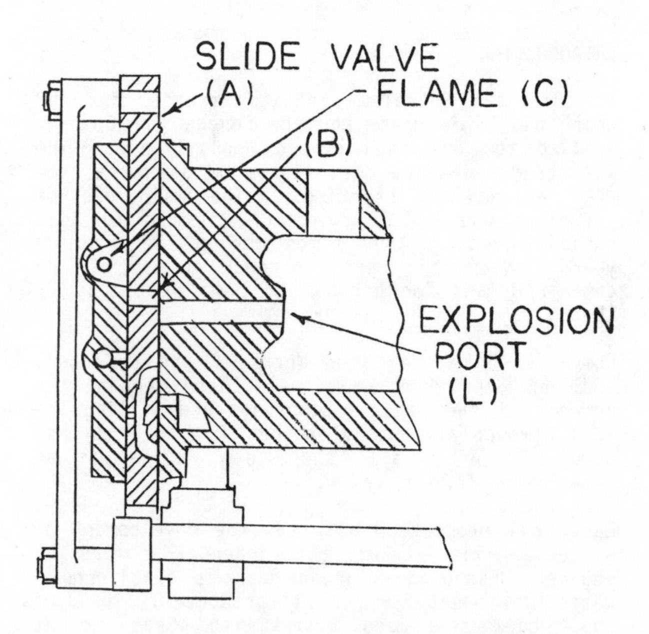

Figure 3
Section View of Cylinder Valve
Arrangement for the Otto Gas Engine

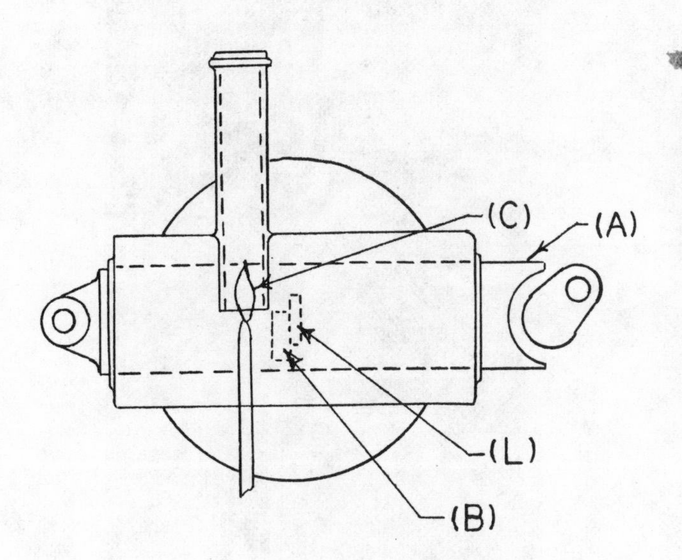

Figure 4
Valve and Ignition Flame for
the Otto Gas Engine

This style of ignition system uses an external flame to ignite the main combustion chamber, which is allowed to burn at all times. As the slide valve (A) moves, it alines the external flame with passage B, which has a gas mixture in it. The flame ignites the fuel mixture in passage B. As the valve continues to move, the burning mixture in passage B lines up with the explosion port L. The burning mixture in B in turn ignites the main combustion mixture through port L. This method allow the igniting flame at B to be extinguished by the explosion, while the main flame C stays burning. Flame C then can rekindle the ignition in passage B in time for the succeeding explosion.

The Superior Gas Engine Company (Cooper Industries, Ajax-Superior), Springfield, O.

The Superior Gas Engine Company was founded in 1889 by Patrick J. Shouvlin in Springfield, Ohio. His immediate objective was to put engineering and machining skills to work servicing the existing oil industries in the area. During this period of time several companies were appearing, particularly on the eastern seaboard, as internal combustion challenged steam as a source of power. Probably the best known among these was The Otto Engine Works of Philadelphia. Many other people were also developing models of their own, and P. J. Shouvlin became one of them, hoping to utilize escaping wellhead gas to power oil-pumping engines.

By the early 1890's, the Shouvlin Machine Shop manufactured its first gas engine. The engine was a 4-cycle design utilizing an open crankcase and hot tube ignition. The engine was also designed with a throttling governor which controlled air and fuel. Fuel for the engine was wellhead natural gas. The engine was also equipped with a reversing clutch. He sold his first gas engine for oilfield pumping to a Findlay, Ohio organization, which later became the Ohio Oil Company, and then Marathon Oil. That year the Shouvlin machine shop became known as the Superior Gas Engine Company. Superior, with its engines adapted particularly for oil fields, proved to have a strategic market location.

As demand continued for engines, a need for service and distribution was required. This was met by the National Supply Company, which became exclusive agents for Superior sales in the oil production industry. Straining limitations of the original machine shop by 1900, Superior Gas Engine Company moved to a new location. Four years later, it moved to its present location.

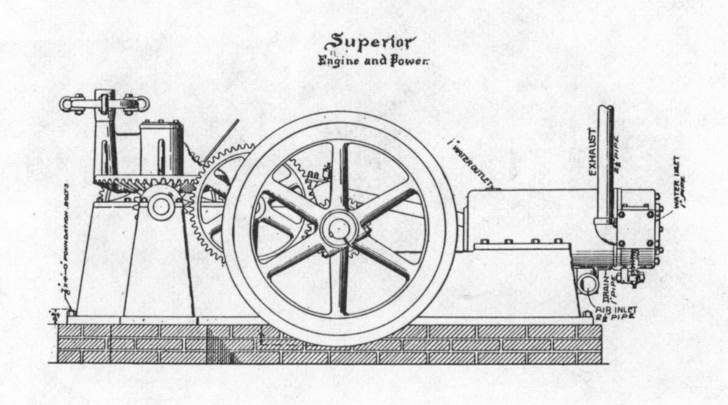

Figure 5
Superior Engine and Power of 1900

By its 25th anniversary in 1914, Superior was selling a line of standard tank-cooled engines from 10 to 100 horsepower. They had expanded from 4-cycle to both 4 and 2-cycle engines. The 4-cycle engines were built in size ranges from 10 to 40 horsepower with an open crank style. Two-cycle engines started at 20 horsepower to 35 horsepower also within open crank style. By this time Superior engines were also equipped with a spark ignition instead of the hot tube.

Superior also introduced a series of engines designed especially to compress natural gas. This style of engines coupled compressing cylinders directly to the engine crank throw.

Figure 6
Typical 4-Cycle Superior Engine of the 1920's

Figure 7
Typical 2-Cycle Superior Engine and
Reversing Clutch of the 1920's

Figure 8
Integral Engine and Compressor
Manufactured by Superior

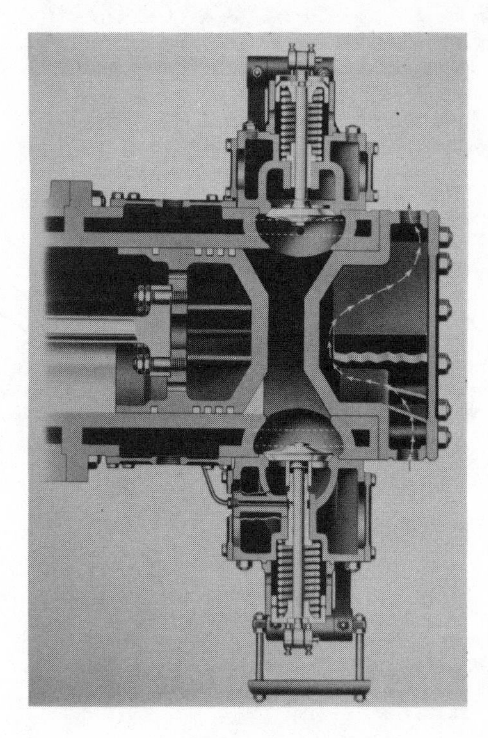

Figure 9
Section View Showing Cross Flow Configuration

The Superior 4-cycle design engine utilized the concept of cross flow head. The intake and exhaust valves were located so that the inlet air and fuel was delivered on one side of the piston and the exhaust was expelled out the other side.

The Superior Gas Engine Company is now a part of Cooper Industries, Ajax-Superior division and manufactures gas engines from 40 to 2700 HP.

Ajax Iron Works, Corry, PA.

In 1877, Mr. L. L. Bliss and C. H. Bagley produced a 12 HP steam engine and formed the company of Harmon, Gibbs & Company. As oil wells grew deeper, the young company incrased the size of its engines, reaching 150 HP. Along with steam engines, the business was supplemented with a general line of machine and foundry work. In 1892, the company was incorporated as the Ajax Iron Works.

Ajax business grew rapidly at the turn of the twentieth century. In 1895, Ajax added gas engines to its existing line of steam engines. By 1911, more than twelve thousand Ajax egnines (steam and gas) had been sold.

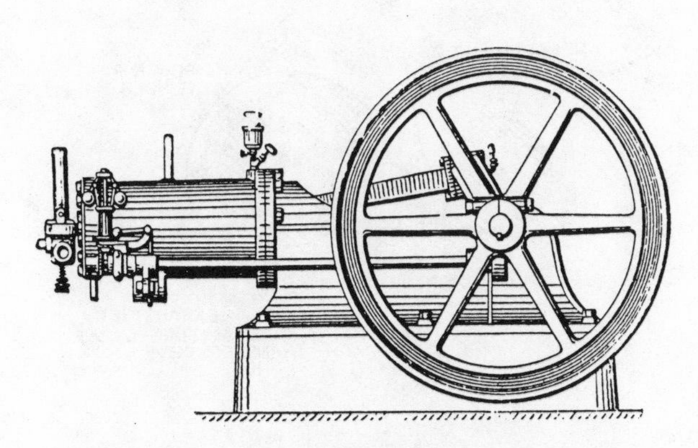

Figure 10
Typical Ajax Gas Engine of 1900

The early gas Ajax engines were very similar to engines of its time. For most oil field engines, ignition was by a hot tube. As with many oil field engines, a crosshead was used, thus eliminating piston side thrust. A side shaft or lay shaft was used to drive governor and gas injection valves.

By 1905, the Ajax gas engine line had grown to include sizes ranging from 5 to 65 HP in a single and 60 to 120 HP in the tandem model.

Although Ajax produced a line of gas engines, its main thrust was developing steam engines. It continued to manufacture steam engine designs through World War II. Today, Ajax is part of Ajax-Superior Engine Division of Cooper Industries.

Bessemer Gas Engine Company, Grove City, PA.

Not all engine companies started out making complete engines. During the period of 1890 to 1900 many of the oil producers were using extremely inefficient steam engines and oil boilers. In analyzing the plight of oil well drillers and producers, it was evident most could not affort to abandon their steam engines and purchase new gas engines. An intermediate step was needed. Dr. Fithian, founder of the Bessemer Gas Engine Company, pondered the idea of replacing steam engine cylinders with those that could be powered by escaping wellhead gas. The idea was sensational but such a changeover necessitated a friction clutch be added. Because there was none to be found, a decision was to make the clutch and cylinders themselves.

In 1898, the Carruthers-Fithian Clutch Company was formed. The company was located in Grove City, Pennsylvania. The Carrathers-Fithian two-cycle gas cylinder and friction clutch, which adjusted automatically to load changes were sold in tandem. For $125, an oil producer could remove a steam engine's cylinder and replace it with a 10 HP gas cylinder-clutch combination.

As demand continued, the company itself took a natural step into gas engine manufacturing. In 1899, the Bessemer Gas Engine Company was incorporated from the Carrathers-Fithian Clutch Company. The company began with a two-cycle 5 HP natural gas engine, and enlarged its line steadily.

The early Bessemer style of engines was horizontal gas engines using a crosshead design. In 1912, Bessemer introduced their commerical engines extensively used in oil field applications. The crosshead design lengthened the engine somewhat over conventional designs. This was a distinct advantage for the crosshead gave total elimination of side thrust on the piston. This greatly prolonged engine life, an essential for the non-stop runs in oil field service. These engines were made in size ranges from 15 to 345 HP.

Figure 11
A 22.5 BHP Bessemer Gas Engine
With Hot Tube Ignition

All of the early Bessemer designs were of 2-cycle natural gas configurations. Most of these engines used a hot tube ignition system and incorporated a throttling governor.

The hot tube ignition is the most common style of ignition system used on early oil field engines. The hot tube ignition uses an external flame to keep a combustion tube red hot. The ignition tube is open to the main combustion chamber. As the engine rotates, the fuel-air mixture is forced by compression to enter into the ignition tube. Since the tube is red hot, the fuel mixture is ignited and continues to burn into the main chamber.

Since most of the engines used in the oil fields use natural gas as fuel, the hot tube system was ideal.

Bessemer also introduced a line of oil engines also. The diesel or oil engines also used the crosshead design. Many of the early Bessemer oil engines were generally interchangeable from gas to oil or vice versa. To those in natural gas districts, this possibility saved a great deal of money in case the gas supply failed. The two fueled units (gas/oil) were equipped with a magneto for ignition.

On April 4, 1929, C. & G. Cooper, and the Bessemer Gas Engine Company merged to form Cooper-Bessemer, the largest builder of gas engines and compressors in America. Cooper-Bessemer continues today as Cooper Industries, Energy Services Group.

Joseph Reid Gas Engine Company, Oil City, PA.

Probably one of the most unique gas engines used in the petroleum industry was the Reid type-A 2-cycle engine. It was designed around the Clerk-cycle engine invented by Dugald Clerk. A distinguishing characteristic of a Clerk-cycle engine was a separate displacer, or pumping cylinder, to provide low pressure scavenging and charging of the power cylinder. The Reid engine was the only commercially American built engine to use this design.

On July 12, 1898, the Joseph Reid Gas Engine Company was issued patent 607276 for the improved gas engine. The Reid engines were designed especially for oil well service. The Reid engine featured hot tube ignition with wellhead (natural) gas as fuel. Most were equipped with the reverse-motion clutch drive needed for drilling and cleaning out wells. An option to the standard engine was a spark ignition and gasoline fueled.

The Joseph Reid Gas Engine Company manufactured the type-A 2-cycle engine in sizes ranging from 4 horsepower to 40 horsepower. The most popular engines, however, were the 15 and 20 horsepower models. The A engine was only made as a single power cylinder engine, but was made in either a left hand or right hand drive. A novel arrangement for the Reid engine was to couple a right hand and left hand engine together, resulting in a configuration similar to a two-cylinder inline piston engine. (See Figure 18).

Figure 12: A Typical Type-A Reid 2-Cycle Engine
This is a 12 HP Engine Built About 1915

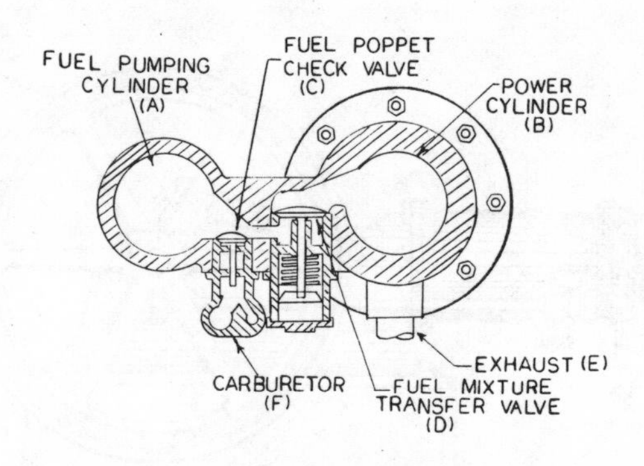

Figure 13
Front View of Reid Engine
Showing Valve Arrangement

A distinguishing feature of the type-A Reid engine was its separate displacer, or fuel pumping cylinder, to provide low pressure scavenging to charge the power cylinder. The fuel pump cylinder piston leads the power piston by 90°. The type-A 2-cycle Reid gas engine works as follows: At position 1, the fuel pumping piston is at TDC and the power piston is 90° BTDC on compression stroke. As the engine continues to rotate the fuel pumping piston draws in a mixture of fuel and air, while the power piston is compressing a fuel air mixture.

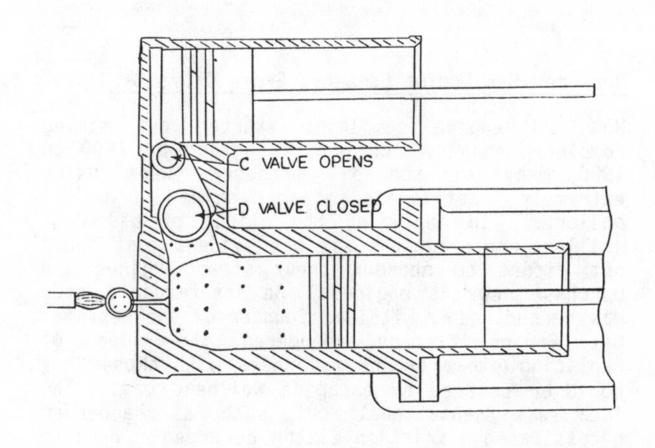

Figure 14:
Postion 1: Fuel Piston at TDC,
Main Piston at 90° BTDC

At position 2, the power piston is at TDC and the fuel pumping piston is at 90° ATDC. At this point some of the fuel air mixture in the main chamber has been pushed up into the hot tube. The tube being heated red hot by an external flame ignites the mixture, and then ignites the main chamber. The rapid pressure rise moves the power piston (giving the power stroke) and the fuel piston is still drawing in the next air fuel change.

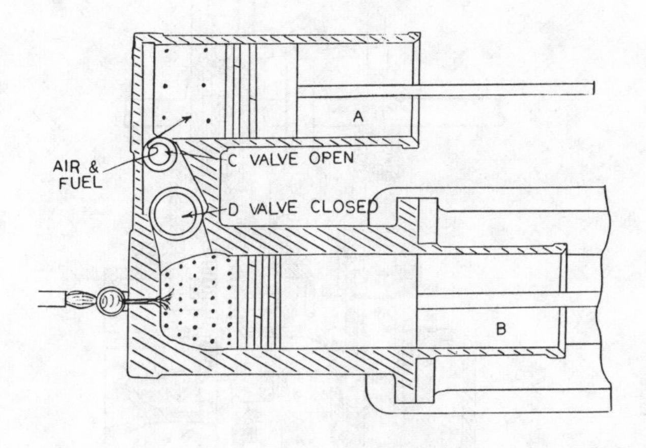

Figure 15:
Position 2: Fuel Piston at 90° ATDC, Main Piston at TDC

At position 3, the main chamber is at 90° ATDC and the fuel piston is at BDC. At this point the fuel poppet check valve closes, and the fuel piston starts to compress the air fuel mixture in the scavenging chamber.

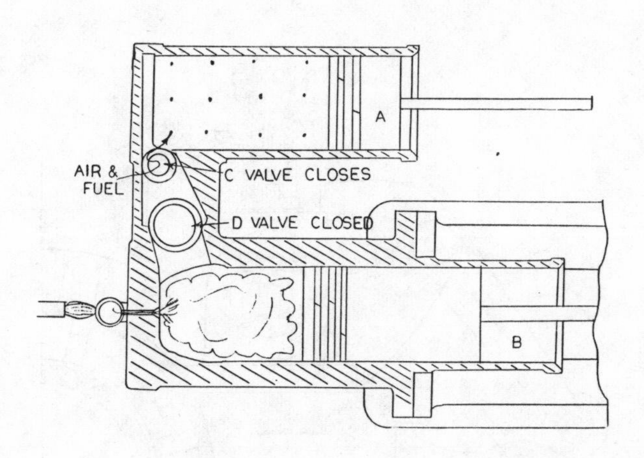

Figure 16:
Position 3: Fuel Piston at BDC and Main Piston at 90° ATDC

At postion 4, the power piston is at BDC and the fuel piston is at 90° BTDC. At this position, the power piston has uncovered the exhaust ports allowing the spent combustion products to exit. With the lower pressure in the main chamber, the fuel transfer valve opens and gas mixture from the fuel pumping cylinder enters the top of the main chamber. This uniflow direction helps scavenge the main chamber. As the engine continues to rotate, the exhaust ports close by the power piston and fuel mixture is compressed in the main chamber. When the presure in the main chamber equals the pressure in the fuel pumping chamber, the fuel transfer valves closes. At this point we are approaching Position 1.

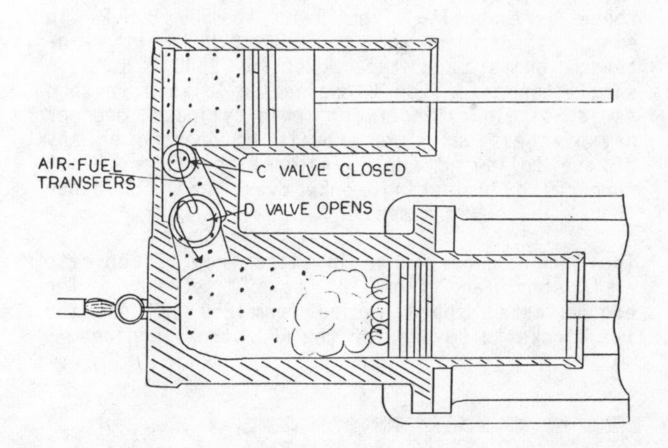

Figure 17
Position 4: Fuel Piston at 90° BTDC and Main Piston at BDC

Figure 18
Left and Right Hand Engines Coupled Together

The Reid Engine Company also introduced to the oil fields a 2-cycle oil engine during the '20's. These engines were made in size ranges from 12 to 40 horsepowers with speeds of 200 to 240 RPM. Late in its career, the Reid company marketed a four-cycle gas engine intended for oil field service. Reid continued in operation until the mid '30's.

Figure 19
Snow Engine, Type B

The Snow Steam Pump Works, Buffalo, N. Y.

This company, the first in the United States to undertake the building of large gas engines, confined its product exclusively to the double-acting 4-cycle horizontal engine. Two types of these were built, Type B up to 500 B.H.P. in single tandem and up to 1000 B.H.P. in twin-tandem units, and type A up to 2500 B.H.P. in single-tandem. Type B engine was also furnished as a single-cylinder or twin-cylinder engine. Any of these units were built to operate on any of the following fuels: natural, producer, blast furnace, illuminating, coke over, and all other industrial power gases.

The Snow engines range in sizes from 11" bore to a 43" bore and from 12" to 60" stroke. The engine rated speed varied from 250 RPM for the 12" stroke to 90 RPM for the 60" stroke engines.

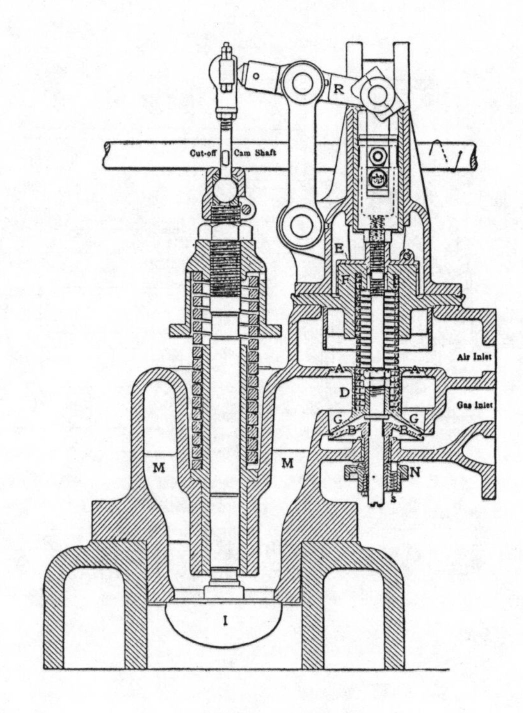

Figure 21
Gas & Fuel Inlet
(Snow Engine)

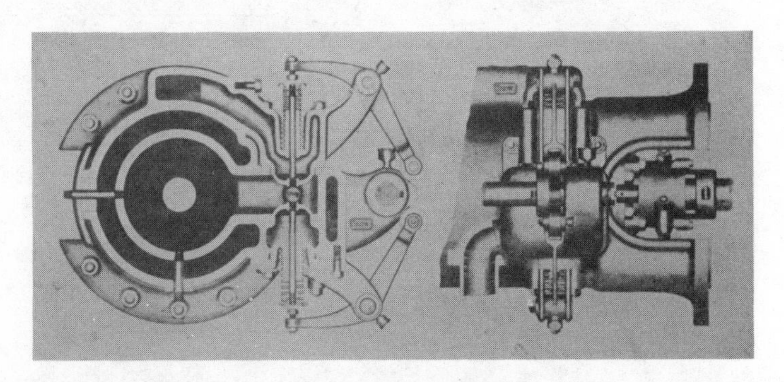

Figure 20
Valve Gear, Type B Snow Engines
200 to 1000 H. P.

There are two features in which the design of both types of engines differs radically from conventional practice. The first is the adoption of the side crank in place of the center crank frame, and the second that the valves all open into a chamber at the side of the cylinder. All inlet valves and their gearing are placed on top of these chambers and all exhaust valves and gearing on the bottom. The great advantage of this construction is that no part of the valve gear is below the floor, and the center of the cylinders can be kept low, making the engine rigid and steady. With this construction only one cam is used to operate both inlet and exhaust valves.

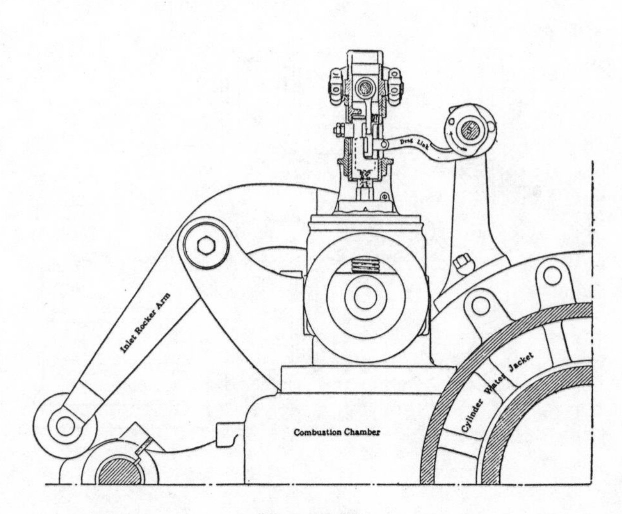

Figure 22
Inlet Rocker Arm Mechanism

The governor on the Snow engine controls the inlet valve cut-off which in turn controls speed. A description of this operation is as follows: In figure #21, the air is admitted through A and the gas through G to the mixing chamber M above the main inlet valve I. Disks A and G, together with the short barrel D, form one casting. The disk G is provided with a taper seat and the length of the valve stem is adjusted in the block at the upper end so that both A and G will seat at the same time. The ratio of air to gas in the mixture is set once for all conditions by adjusting the baffle disk B by means of the knurled nut N. The cut-off valve is operated as follows: Inlet valve I is opened by the rocker arm, Figure 22, always at the beginning and end of the suction stroke. By means of the link R, Figure #21, the main inlet valve stem operates a sliding block in a guide above the cut-off valve. To this sliding block is pivoted a latch, L, Figure 21, which, when in position, engages with the block on the end of the cut-off stem and thus lifts this valve when the inlet valve I opens. To close the cut-off valve the governor controls the latch L, dragging it out of position by means of the drag link and the cam C shown in Figure 22. The right hand end of the drag link curves around the govenror shaft S and rests upon the journal box, thus holding the link in place as it slides back and forth. The cam C engages a lug on this link, thus displacing it and the latch L. At the moment the latter frees the block, the cut-off valve is closed by means of its spring, Figure 22, and no more mixture is admitted to the cylinder. The cam shaft S rotates continuously at half speed. The point in the suction stroke at which cut-off occurs depends entirely upon the position that the cam, C, with relation to the crankshaft, and this is controlled by the governor through a so-called "floating" bevel gear.

The Snow Steam Pump Works became part of Worthington Pump and Machinery Company in 1918.

C & G Cooper, Mt. Vernon, O.

In 1833, Charles and Elias Cooper opened a foundry in Mt. Vernon appropriately named the Mt. Vernon Iron Works. To improve their foundry blast furnace in the fall of 1836, the Coopers built a steam engine to run the fans. The now know C & E Cooper started to manufacture single cylinder, slide-valve blower engines for blast furnaces. In 1853, the newly known C & J Cooper Company built a wood-burning locomotive, the first manufactured west of the Allegheny Mountains. By the time of our Civil War (1860), Cooper was manufacturing stationary steam engines for oil field service. By 1869, Cooper started manufacturing large Corliss steam engines.

By 1895, the Cooper firm went through another name change being known now as C & G Cooper Company. As natural gas and oil sources were being discovered, the internal combustion engine was challenging the King steam engines. Large engines fueled by natural gas were proving successful in the mushrooming steel and oil industry. A new need for gas compression and pipeline transmission had emerged.

In June of 1909, the company's first gas engine-compressor combination was installed. The unit was a 20 inch by 42 inch twin tandem gas engine rated 730 BHP at 90 RPM for compressing natural gas.

Cooper's big horizontal, 4-cycle, double-acting gas engines were commonly used after 1913. In 1919-1920 a line of small horizontal, 4-cycle, single-acting gas engines of about 80-160 BHP were introduced. The smaller single-acting engines together with a line of compressor cylinders immediately became popular in smaller gas pumping stations. Hundreds of the Cooper-direct-drive compressor units were sold between 1920 thru 1929 to compressor plants for the manufacturer of natural gasoline.

In 1929, the C and G Cooper Co. and Bessemer Gas Engine Co. merged to form Cooper-Bessemer Corporation. Cooper-Bessemer would later become Cooper Industries, a multi-billion dollar corporation.

SUMMARY:

There are many more engine manufaturers of oil field engines too numerous to mention. Springfield, Ohio for example had a total of 13 different engine manufacturers during the last of the the nineteenth and first quarter of the twentieth centuries. I have tried to pick the more unusual designs or well-known manufacturers that were producing engines prior to 1900. Since then many more engine designs and manufacturing evolved, most have gone, a few of the better ones remain.

Today new technology has produced many changes to the gas engine. High energy spark ignition, pre-chamber-lean combustion, high speed, and multi-cylinders have all advanced the engine industry.

General References:

1. Cummins, Jr., C. L., "Internal Fire", Carnot Press, 1976.

2. Guldner, H. and Diederich, H., "The Design and Construction of Internal Combustion Engines," D. Van Nostrand Co., 1910.

3. Joseph Reid Gas Engine Company Patents 984023 issued 1911 and 607276 issued 1898.

4. Keller, D. N., "Cooper Industries 1833-1983," Ohio University Press, 1983.

5. Setright, L. J. K., "Some Unusual Engines," Mechanical Engineering Publications Limited, England, 1975.

6. Warwick, P. B., "The Gas Engine How it Works and How it is Used," Bubier Publishing Co., 1907.

7. Wendel, C. H., "American Gasoline Engines Since 1872," Crestline Publishing Co., 1983.

DIESEL ENGINES OF 2,000 BHP PER CYLINDER – PRE-1914 MARINE ENGINE DEVELOPMENTS

C. L. Cummins, Jr.
Carnot Press
Wilsonville, Oregon

ABSTRACT

There was a hundred fold increase in power output per cylinder between Rudolf Diesel's 1897 "Rational Heat Engine" and its progeny of less than twenty years later. Marine diesels were already propelling ocean-going merchant ships at great cost savings over steam, and its adaptation to the submarine made this lethal weapons system practical for the first time. Discussed are European two and four stroke cycle marine diesel engine and related fuel system developments between 1903 and 1914. Emphasized are the 2,000 bhp (1,491 kW) per cylinder German and Swiss test engines intended for battleships.

INTRODUCTION

A proud inventor showed off his 20 hp (14.9 kW) lab engine to potential licensees in February 1897 after a grueling four year development struggle at M.A.N. in Augsburg. But Rudolf Diesel's dream of easy wealth turned to despair when his first engines went into service a year later. It was all he and a few dedicated helpers could do to keep customers from returning them. Licensees who had invested large sums were doubly discouraged; a few had their up-front fees refunded. The worst problems included nozzle carboning and exploding air injection compressors. Not until after 1900 did skeptical customers begin again to place significant orders.

Several licensees judiciously used this interim for further development and sold no engines. Some also strayed from the designs that Diesel had furnished and soon equalled or surpassed the performance achieved in Augsburg. These were still four-stroke, single or two cylinder engines of under 50 bhp (37 kW).

FUELS

Although it was claimed that the diesel engine ran on almost any liquid hydrocarbon, the fuels of choice were distillates similar to Nos. 2 to 4.[1] Heavier fuels (not today's residuals) were sometimes used when preheated. Diesel fuels were between the kerosene lamp and lubricating oil products from the distillation process used prior to thermal cracking. Since crude stocks varied from field to field in sulfur and asphalt content, the engine builders often specified fuel for their diesels by where it came out of the ground. One percent sulfur was considered acceptable and ways were found to tolerate even higher amounts.

Taxes played an important role in the fuels for diesel engines. Countries with little or no crude oil resources often imposed import duties equalling the base fuel cost. This raised user prices so high that the diesel was not competitive with steam. In countries having high import duties the engines ran on domestically manufactured fuels made from local oil shale, coal tars and even lignite. Most of Western Europe, where the engine was trying to make inroads, imposed these taxes. Diesel powered merchant ships with their long cruising ranges, however, could resupply in those ports where fuel costs were low.

FUEL INJECTION

Early engines used air ("blast") fuel injection. Diesel grudgingly adopted it because he could not create high enough hydraulic pressures to meter and atomize a fuel charge by a pump and nozzle made with known technology.[2] Yet air injection meant adding a costly and cumbersome, single-stage compressor that robbed up to ten percent from indicated power. Most compressors were engine driven to supply both starting and injection air. The mentioned compressor explosions were caused by a too hot discharge air and lubricating oil from the compressor forming a combustible mixture. Multi-staging with intercooling and better oil separation cured this.

These diesels had compression pressures of 450-500 psi (31-34.5 bar), with peak pressures of about 100 psi (6.9 bar) above that. Full load injection air pressure was between 800-1,000 psi (55-69 bar).[3]

The injection cycle began by depositing a fuel charge in the nozzle's lower end from a plunger pump whose output was changed by either throttling the pump suction or the discharge. Fuel pressure was only slightly higher than the injection air pressure in the nozzle. Between the fuel port and the seat of a cam-operated nozzle valve was an atomizing sleeve around the valve's stem. When the cam lifted the valve just before TDC, the fuel and oil were blasted through the atomizer to create a very fine mixture that was injected into the combustion chamber through nozzle holes. Under full power, injection continued as long as the cam held open the valve. The cam also determined the rate of admission so that peak pressure rose little above compression pressure to give Diesel's classic "constant pressure" indicator diagram.

The major builders developed their own atomizers -- either a stack of serrated washers, a notched sleeve, or a venturi sleeve -- to reduce smoke and nozzle hole carboning. Although air injection prevailed prior to World War I, Vickers perfected a common rail, "solid" injection system for their submarine diesels after 1912. They had spent years reaching this point and applied for patents, beginning in 1910, under engineering director James McKechnie's name. Little was known of this British system, except for the patent literature, until after 1918. The design entering service used either four or six plunger pumps packaged together, the number depending on engine size. These discharged into a manifold leading to the nozzles. The maximum fuel pressure leaving the pump was about 4,000 psi (276 bar). For load control, this pressure was modulated by an operator and governor controlled valve at the pump exit. Later systems controlled the suction valve on the inlet side of the pump. A cam-operated needle valve in the nozzle timed the injection and its duration.[4]

One other "airless" system, the Brons from Holland (with modifications patented in the U.S. by Hvid), was used in a few small marine compression ignition engines starting around 1905. A distinguishing characteristic was its small antechamber into which the fuel charge was deposited under low pressure during the intake stroke. The charge ignited when heated air at the end of the compression stroke entered the antechamber through passages connecting it with the main cylinder.

RAIL AND ROAD DIESELS

Locomotive and truck diesels before 1914 can be seen only as first attempts. The most successful example was by Sulzer who in cooperation with Diesel built a locomotive engine for trial service with the Prussian State Railway in early 1913.[5] It was a crosswise, ninety degree V-4 diesel having a bore and stroke of 380 x 550 mm. The reversible 2-stroke engine had a nominal rating of 1,000 hp (746 kW) at 304 rpm using a nominal bmep of 90 psi (6.2 bar); a peak power of 1,600 bhp (1,193 kW) at 177 psi (8 bar) bmep was claimed possible for brief periods. The two drive axles were linked by tie rods, as in steam locomotive practice, to an overhung crankpin on the flat outer face of a flywheel at each end of the engine. A separate 250 hp (186 kW) diesel drove the compressor to maintain pressure in the starting air tanks. Because of the permanent link with the wheels, air also started the train as it pushed down the pistons.

Diesel himself sponsored a 40 bhp (30 kW) truck engine with solid fuel injection in 1910, but it smoked so badly on the test stand that further effort was cancelled.[6] The Nürnberg division of M.A.N. also designed a family of automotive/rail engines which never left the factory.[7]

EARLY MARINE DIESEL ENGINES

Larger marine diesels required a greater structural stiffness. The typical land based A-frame design, with cylinder and pedestal bolted to a bed plate holding the crankshaft, was adequate when all rested on a rigid subbase. But a ship's hull flexing in a heavy sea was not the same. Marine steam engine practice was heeded, but the added cylinders and higher peak bearing loads required deeper bed plates and A-frames strongly tied together. Pressure lubrication and water cooled pistons also called for enclosed crankcases, especially on a rolling and pitching ship. This led to cast or plate box sections.

Height-limited submarine engines went to larger cylinder bores that sometimes equalled the strokes. Engines were also lowered by using a trunk piston to eliminate crosshead and piston rod, albeit making for a longer piston.

Pre-1914 marine diesels evolved in stages:

1) Non-reversible 4-stroke (1903-04).
2) Two and 4-stroke reversible engines for surface and submarine use (1906-08).
3) Ocean-going merchant ships (1910-12).
4) 2,000 Hp (1,491 kW)/cylinder naval test engines (1912-14).

Fig. 1. Dyckhoff opposed piston diesel engine in the canal boat Petit Pierre, 1903.

The specific weights of marine diesels depended on how conservative the design. Submarine engines varied between 16 to 18 kg/hp, and surface ship engines from 70 to 115 kg/hp or more. Reported weights must be viewed with caution as they may or may not include needed auxiliary equipment such as injection air compressors and cooling pumps.

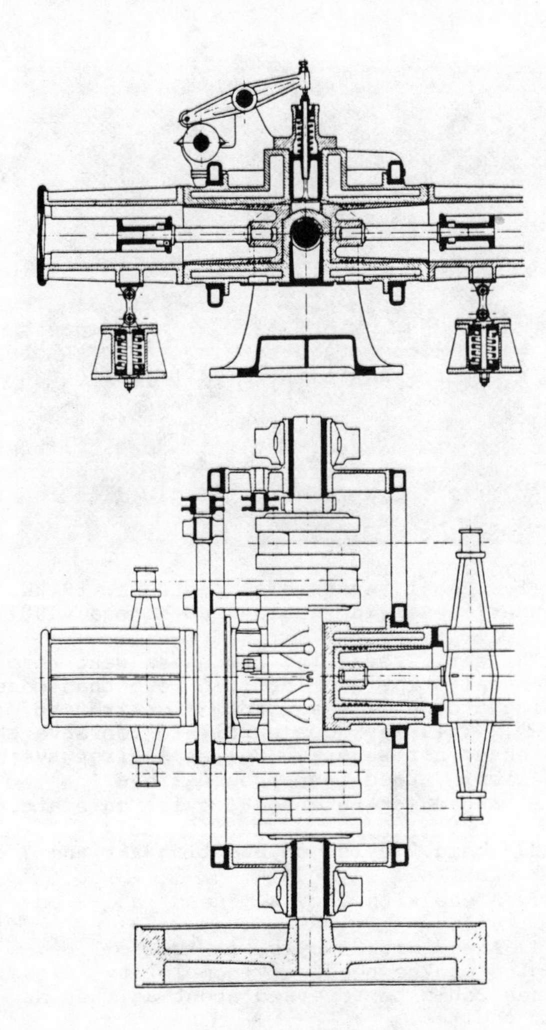

Fig. 2. Dyckhoff 19 kW, opposed piston diesel, 1903. One cylinder, 2 pistons.

Caution must also be applied to published brake specific fuel consumption. For this reason bsfc data are not given unless it is known what is included. However, an average figure of 0.44 lb/bhp hr (268 g/kW hr) can be used for the engines in this period.[8]

The first marine diesels were installed in 1903. One was by F. Dyckhoff Fils in Bar-le-Duc, France and and another by the Stockholm based A/B Diesels Motorer, the company where Jonas Hesselman contributed so much.

The 20 hp (14.9 kW), 4-stroke French engine went in the canal boat Petit Pierre. (Fig. 1) Frédéric Dyckhoff, a long time friend of Diesel's, and Adrien Bouchet designed this unusual 2-piston, inwardly opposed engine. The crankshaft passed through an "alley" to form a partial cylinder head between the two pistons; a passage above the shaft alley joined the coaxial cylinders.[9] (Fig. 2)

A/B Diesels Motorer sold three engines to Nobel Brothers of St. Petersburg (Leningrad) for the Vandal, a new design river tanker bringing kerosene up the Volga from the Caspian Sea. These 3 cylinder, 4-stroke engines each produced 120 bhp (89 kW) at 300 rpm. Bore and stroke were 290 x 430 mm. Nobel built their own successful engines after this.[10] (A/B Diesels Motorer also had two engines in the Great Lakes ore boat Toiler, the first diesel powered ship to cross the Atlantic in 1911 -- at 7 1/2 kts.)

M.A.N. ran a 4 cylinder engine of 140 bhp (104 kW) at 400 rpm that year, but it never entered service. Yet, within three years they installed 300 bhp (224 kW) diesels in French submarines.

REVERSIBLE DIESEL ENGINES

A disadvantage for I-C marine engines, when compared with steam, was their inability to reverse rotation for maneuvering. Available reverse gears and reversible propellors were power limited.

The Petit Pierre used a reversible prop. The triple screw Vandal devised an electric drive: generator, magnetic clutch and motor, all coaxial with the shaft. In "ahead", the clutch engaged to directly couple engine and propeller. In "astern" the clutch was disengaged, and the generator drove an opposite turning motor. This system was like one concurrently made by the Italian engineer Del Proposto. Sulzer Brothers' first marine diesel, adapted from a stationary model and with a Del Proposto system, went in the Lake Geneva cargo

Fig. 3 Sulzer 4-stroke, 30 kW diesel in the Lake Geneva cargo boat Venoge, 1904.

boat Venoge in 1904. Its 2 cylinder, 4-stroke engine developed 40 bhp (30 kW) at 260 rpm. Bore and stroke were 230 x 350 mm.[11] (Fig. 3)

Dyckhoff built an advanced, reversible locomotive diesel in 1899, based on his German patent 107,395, but torsional vibration problems prevented it from leaving the shop. There were two disengageable camshafts, one for "forward" and the other for "reverse". The 3 cylinder engine, with a square(!) bore and stroke of 130 x 130 mm,[12] was to produce 15 bhp (11.2 kW) at 600 rpm.

Sulzer displayed a reversible diesel at a Milan marine exhibition in 1906. (Fig. 4) The 4 cylinder, 175 x 250 mm engine produced 100 bhp (75 kW) at 400 rpm. It was their first 2-stroke, with uniflow scavenging by two inlet valves in the head and exhaust ports uncovering near BDC.[13] A high mounted camshaft opened an inlet valve in a detachable valve cage set in a monobloc head and cylinder. Through bolts tied these and an intermediate, one piece crankcase to the bedplate. A 2-stage injection air compressor at the end opposite the flywheel was crankshaft driven. The scavenge compressor operated off a rod linked to the high pressure compressor connecting rod. For reversing, the engine was stopped, the rollers lifted from the cams. The camshafts were then rotationally shifted to change starting, injection and valve

Fig. 4. Sulzer reversible 2-stroke, 75 kW marine diesel exhibited in Milan, 1906.

timing, after which the rollers were again lowered. The engine was restarted by air admitted through a starting valve in the head. Another new feature was the absorbing of the propeller reaction by a thrust bearing added at the rear crankshaft main bearing. Before this, external thrust bearings on the propeller shaft were used. The exhibited engine and four of a modified design were installed in river barges during 1909-1910.

Nobel built a 4-stroke reversible engine in 1907; the following year two went in the 117 ton Russian submarine Minoga.[14] (Fig. 5) These 3 cylinder diesels of 275 x 300 mm produced 120 bhp (89 kW) at 400 rpm and propelled the vessel at 8 kts. A family of engines with more and larger cylinder bore and strokes went in subs and other naval ships. The Minoga engines used sliding camshafts. Later ones, with fixed forward and reverse cams, had rocker lever linkage which could be positioned so that either one or the other of two out-of-phase roller cam followers contacted the appropriate cam. M.A.N. also adopted this concept for some of their engines.

Fig. 5. Nobel reversible, 4-stroke, 89 kW, 3 cylinder diesel in Russian sub Minoga, 1907.

The early reversing processes went through sequential steps and involved more than merely turning a control wheel from "forward" to "reverse". This was done partly to conserve the compressed air supply. Sulzer's first system, for example, used four progressions:[15]

1) Ahead (or reverse) 4 cyls. on starting air.
2) Ahead 2 cyls. on starting air and 2 on fuel oil.
3) Ahead with 2 cyls. "dead" and 2 on fuel oil.
4) Ahead with 4 cyls. on fuel oil.

In spite of the necessary complexity, diesel engines could be reversed about as fast as steam engines.

Fig. 6. Sulzer cross-scavenged, 2 stroke, 283 kW diesel in cargo ship Romagna, 1910.

OCEAN-GOING SHIPS

Sulzer went all 2-stroke for marine diesels with their two, 4-cylinder, loop scavenged engines for the Romagna, a 1,000 ton cargo ship launched in 1910. (Fig. 6) The trunk piston engines produced 380 bhp (283 kW) each at 250 rpm from a bore/stroke of 260/350 mm.[16] The cross port scavenging included two rows of intake ports with an upper row adding a degree of supercharging. A cam operated valve opened the upper ports to the scavenge air supply after the exhaust ports had been cut off on the piston's upstroke. (Fig. 7) The lower air ports closed before the exhaust ports. Eliminating poppet inlet valves greatly simplified

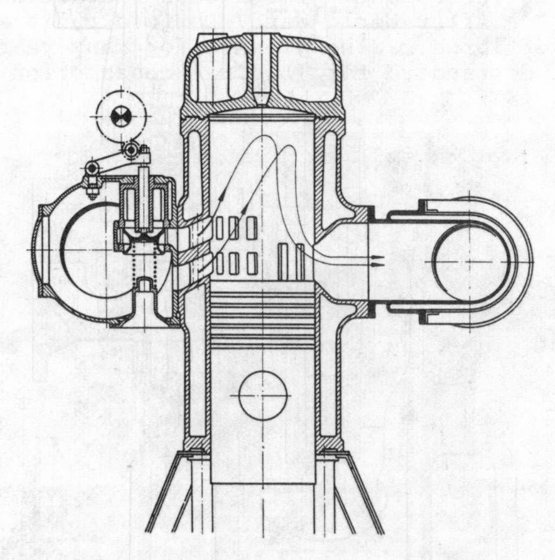

Fig. 7. Sulzer cross flow scavenging with cam actuated valve for upper intake ports, 1909.

the reversing mechanism and reduced cylinder head cracking around the casting openings for the valves. Cracking was a continuing problem for all engine builders. After the vessel had entered service the compressor for air injection and starting was found to inadequate and was replaced. The Romagna sank in February 1911 during a sudden storm in the Adriatic Sea after its cargo had shifted.

Fig. 8. M.S. Selandia, 1912. First major diesel powred ship. Two Burmeister & Wain 4-stroke, 932 kW diesel engines.

Burmeister & Wain of Copenhagen, an early Diesel licensee, began their marine diesel activities with the Selandia, a cargo vessel of 10,000 tons displacement launched in 1912. The absence of a smoke funnel -- the exhaust went up a hollow rear mast -- gave the Selandia a distinctive appearance. (Fig. 8) The two 8 cylinder, 4-stroke engines produced 1,250 ihp (932 kW) each at 140 rpm from a bore and stroke of 530 x 730 mm. (Fig. 9) As was standard on all larger marine diesels for surface vessels,

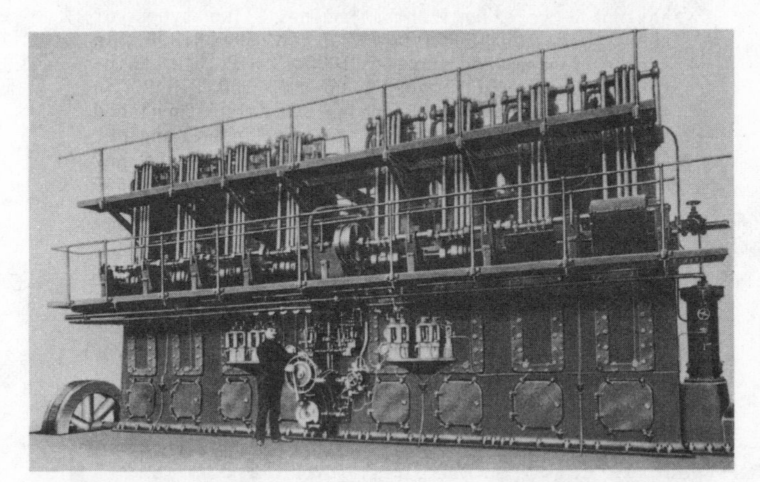

Fig. 9. Burmeister & Wain reversible, 6 cylinder, 4-stroke diesels in the Selandia, 1912. 932 kW at 140 rpm.

these engines were of the crosshead type.[17] Auxiliary compressors, for maneuvering and the first stage of injection air, went to 300 psi (20.7 bar). An engine driven, second stage injection air compressor boosted this to 900 psi (62.9 bar). Brake output was about 1,125 bhp (839 kW) when the power to drive the auxiliary compressor was deducted. Reversing was by axially shifting the camshaft to bring a second set of cams into play. The original engines remained in the Selandia until it went down off Japan in 1942. Significant operating savings helped sell similar sister ships until the 1914 war began. Of equal importance with the lower costs was a 900 ton fuel capacity in double bottoms that gave her a potential cruising range of about 20,000 miles. Harland & Wolff (Belfast) became a B&W licensee in 1913.

Although Krupp had been a full partner with M.A.N. in the original support given Diesel, their engine program was small until their Germaniawerft division began in 1909 to build diesels for their submarines.[18] (Until then all German U-Boats used Körting low compression oil engines for power.)

Krupp's first uniflow 6 cylinder, 2-stroke engines, not operational until 1911, produced 850 bhp (634 kW) at 450 rpm from a bore/stroke of 350 x 400 mm. Of special interest were the three scavenge air valves in the head opened by a three-fingered lever actuated off a single cam. (Fig. 10) The fuel injector, opposite one of the valves, was inclined at 45 degrees but with a vertical nozzle hole opening at the cylinder axis.[19]

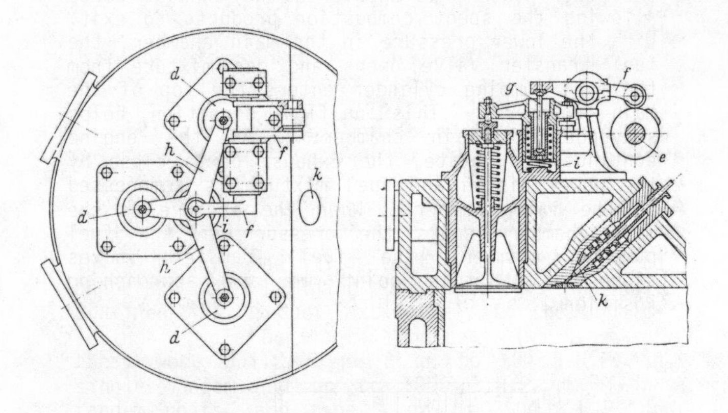

Fig. 10. Krupp 2-stroke, 634 kW submarine engine, 1910, with angled fuel nozzle and 3 intake valves for uniflow scavenging.

By 1914 Germaniawerft was building 1,100 bhp (820 kW), 2-stroke sub engines with a bore/stroke of 390 x 450 mm. Rated speed was 390 rpm. They were still uniflow scavenged with only one inlet valve in the head. A larger version in 1916 produced 1,650 bhp (1,230 kW) at 350 rpm. The starting and injection air compressors (a four stage for the latter) made up almost one third the total engine length.

Krupp built a few 4-stroke diesels that went in cargo submarines. With a bore and stroke of 320 x 420 mm, the engines had an output of 450 bhp at 400 rpm. One of these, the Deutschland, made a round trip to Baltimore from Kiel in 1916 which created headlines in the papers and further demonstrated the cruising range potential of underwater craft.[20]

Two-stroke merchant ship engines were also built by Krupp at Kiel. The 6-cylinder crosshead design, rated at 1,100 bhp (820 kW) at 120 rpm, had a bore and stroke of 450 x 800 mm. The pistons were water-cooled by articulated tubular linkage to the crosshead and piston rod that supplied water to and carried it away from the piston (common practice on several makes).

Carels of Belgium, one of the first Diesel licensees, entered the marine market in 1909 with a trunk piston, 2-stroke, 1,000 bhp (746 kW). It was the only larger size engine without crossheads. By 1914 they offered a 6 cylinder engine of 1,800 bhp (1,342 kW) at 100 rpm from a bore and stroke of 600 x 1,100 mm. The first one was built by Reihersteig, a Hamburg licensee, for the single screw 7,960 dwt tanker Wotan which performed well in Atlantic service.[21] The Carels marine diesel delivered the highest output of any yet entering service.

Professor Hugo Junkers merits mention in any diesel history of this period. His 1892 inwardly opposed, uniflow scavenged 2-stroke gas engine was adapted to a diesel. A lengthy test program was carried out in his Aachen lab, but all resulting commercial engines were built under license. Junkers vertical diesels, with their intake and exhaust ports at opposite ends of the cylinder, went in two merchant ships. The first, in 1912, were a pair of 850 bhp (634 kW), 3 cylinder "tandem" engines by Weser A.G. for the Primus. (Fig. 11) Crank throw and rod extensions operated an upper pair of pistons in a coaxial cylinder. Thus the engine had four pistons per each crank throw. (Fig. 12) For reasons of failure or owner concern the engines were replaced by steam before the vessel left the dock.[22] The second pair installed in the Arthur von Gwinner fared better. These non-tandem 2 cylinder, 4 piston diesels of 850 bhp (634 kW) had a bore and stroke of 440 x 520 mm. Highest reported output was 814 bhp (607 kW) at 185 rpm with a mechanical efficiency of 74 percent. Bsfc at that rating was .43 lb/bhp hr (263 g/kW h.).[23]

This ended Junkers work on big marine diesels, but Doxford, the British licensee, carried on the opposed piston design starting in 1913.[24] After World War I Doxford built a line of large marine diesels for many years and set the standard for low fuel consumption.

Fig. 11. Junkers-Weser 3 cylinder, tandem, opposed piston (12 pistons), 634 kW, diesel for the cargo ship Primus, 1912.

Not all pre-1914 marine engines can be covered. However, several companies, in addition to those already mentioned should be cited as having advanced the position of marine diesels. A 500 bhp (373 kW), 4-stroke, 6 cylinder Werkspoor diesel powered the ocean going tanker Vulcanus in 1910. The Italian navy bought a Werkspoor 2-stroke sub engine in 1911.[25] FIAT began their 2-stroke submarine diesel program in 1909. Sautter-Harlé of France the builders of Dyckhoff's Petite Pierre engine, made sub diesels starting in 1907 that competed well with those by M.A.N.

Conspicuously absent are United States companies. Not until almost the War was a major effort devoted to marine diesels, and then most of the first designs came from Europe. Nlseco

(New London Ship and Engine Co., a division of Electric Boat Co.) bought a license from M.A.N. for the first U.S. submarine diesels. They also built Vickers diesels for "Holland" type subs destined for England which later travelled there without incident.[26] (Nlseco, who held the Holland patents, had earlier licensed Vickers to build Holland boats.)

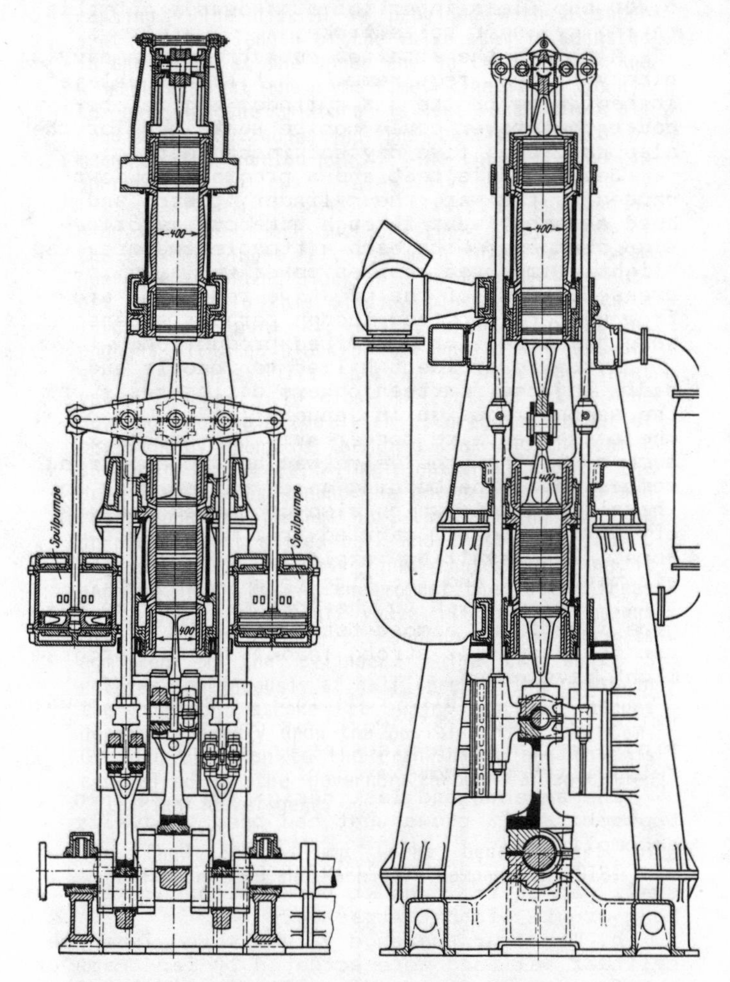

Fig. 12. Junkers-Weser tandem, opposed engine, 1912. Height: 7.25 m.

Diesel and early devotees of his engine had given little credence to the 2-stroke cycle, but this quickly changed as acceptance of marine diesels grew. Even with a recognized fuel consumption penalty over the 4-stroke of about ten percent, the advantages of smaller specific volume and less complex reversing mechanisms carried the day with many builders. Unfortunately, those dedicated to either 4-stroke or 2-stroke carried their strong convictions to such an unhealthy point, with one group downplaying the other, that the diesel user was confused and the engine's image suffered. This conflict climaxed in the 1920's.

A double-acting engines, with a combustion chamber on each side of the piston, was yet another way to raise power at little additional engine room space. When combined with the 2-stroke, the diesel gave a power impulse on every piston stroke. Double-acting did not come easily due to problems with scavenging, piston rod sealing and piston and cylinder wall cooling. Burmeister & Wain introduced 4-stroke double-acting engines in the 1920's.

DIESEL ENGINES FOR BATTLESHIPS

European navies observed the improving reliability and rapidly increasing power of merchant marine diesels. Only Germany acted. In 1909 the Nürnberg plant of M.A.N. suggested to the Reichsmarine that they were prepared to build a diesel engines capable of powering a capital ship. Within months a development contract was signed for a 160 rpm, 6 cylinder, 12,000 bhp (8,950 kW) battleship engine.[27] Six of these, driving three shafts, would provide the 70,000 hp (52,200 kW) needed. Krupp's Germaniawerft factory received a similar contract. Both were to demonstrate an output of 2,000 bhp (1,490 kW) per cylinder in a three and then a six cylinder engine for installation in a ship. Each cylinder's output exceeded the total power of the largest six cylinder marine diesel in operation when the program began.

Sulzer, also wanting to compete, only built a single cylinder test engine as they were to fund it themselves.

All three companies chose 2-stroke. Krupp and M.A.N. opted for double-acting; Sulzer's would be single-acting. M.A.N. and Krupp used inlet valves for uniflow scavenging; Sulzer had ported, cross flow scavenging. The bore and stroke of the single-acting Sulzer were the largest:

Sulzer: 1,000 x 1,100 mm
Krupp: 875 x 1,050 mm.
M.A.N.: 850 x 1,050 mm

None of the Groß-Dieselmaschinen escaped thermal stress related difficulties which took years to overcome. Piston, cylinder walls and head all suffered cracking or outright failure, and only through persistent, creative redesign was progress made. Mechanical weaknesses were less troublesome, although at times disastrous.

THE "NÜRNBERG" ENGINE[28]

Although Nürnberg's previously largest diesel was only 100 hp per cylinder, they could draw on a long experience of building huge horizontal gas engines. Anton Rieppel, the talented company manager who dared to promote the giant engine in the first place, also provided the guidance and courage necessary to complete the project.

One must appreciate the immensity of these earliest "cathedral" diesels: The height of the M.A.N. engine was over 7.4 meters. Its cylinder and head structure, the area causing the most grief, had an overall diameter of about 1.5 by 3.3 meters high.

The base plate and crankcase structure supporting the cylinders were of steel. A double-sided crosshead guided the piston rod and held the upper connecting rod bearing. (Fig. 13) Lubricating oil cooled a hollow, two-piece piston. The oil supply passed from the crosshead and up through an axial tube in the hollow piston rod.

The upper and lower heads of cast steel were bolted to a single piece cylinder containing the exhaust ports. The cylinder heads each held four scavenge valves, four fuel valves and two valves combining the air starting and safety popoff. Four camshafts -- two on each side of the engine -- actuated the twenty valves per cylinder assembly.

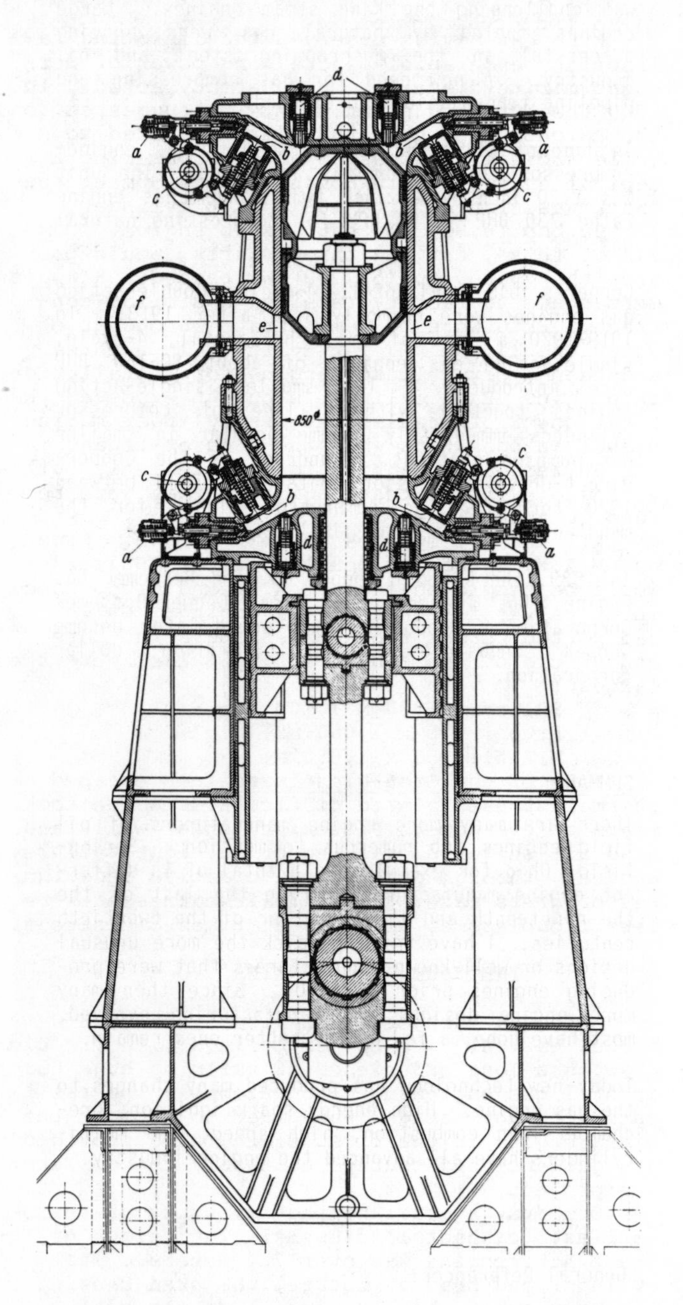

Fig. 13. M.A.N. 3 cylinder, 2-stroke, double-acting engine, 1911. Rating: 1,490 kW/cyl.

Three double-acting scavenge air pumps (enough for the six cylinder engine) were engine driven. Their bore and stroke were 1,320 x 800 mm each, or 45 percent larger in displacement than a power cylinder. High pressure air for the blast injection and starting were driven by a 2-stroke, 350 hp. Nürnberg diesel.

In March 1911, one year from design start, the three cylinder engine was put on test. Results were not encouraging. The scavenge and blast air compressors broke pistons and valves. Of greater import, heat related problems occurred at the power cylinders: Pistons were under cooled and soon cracked; cylinders cracked between the exhaust ports; heads cracked around the valve ports. New heads were of cast iron to expedite replacement parts. Nevertheless, the maximum power was briefly increased to 5,400 bhp (less injection air needs). At this rating cylinder bolts broke.

Although the attained output met the navy's ninety percent requirement, M.A.N. was delayed in beginning on the six cylinder engine, because that power could not be sustained for the also specified five day endurance test.

Between this test and a program slowdown caused by the war, the cylinder, piston and head assembly went through numerous major design changes. With each a troublesome area was slightly improved even as power was also increased. Some of the creative redesigns arose from desperation, and though perhaps expensive, were practical for a limited production.

A tragic accident killed ten people and badly injured fourteen others during tests of the second redesign in January 1912. One of the scavenge valve rocker arms had broken in such a way that the valve was held open during combustion. The burning gases passed back into the air manifold which also contained pockets of lubricating oil carried from the compressors. The resulting explosion blew off the air manifolds and set fire to a fuel supply tank. In addition to the loss of life the engine itself was almost totally destroyed. If not for Rieppel's strong leadership the program might have ended.

Tests on a third redesign did not begin until March 1913. All work was confined to a single cylinder, a practice in effect until reliability was more evident.

The seventh and last design (Fig. 14) in September 1913 shows what had been gradually accomplished:

a) A two-piece cylinder separated by a small gap at the exhaust ports allowed axial heat growth without stresses.

b) The four scavenge valves moved into the cylinder wall and were actuated by face cams on the four camshafts. (Fig. 15) Two fuel valves instead of four per chamber also moved to the cylinder wall. The upper head held one starting valve; none was in the lower head.

c) A spiral cooling rib directed water flow at high velocity along the wall area transferring the most heat.

d) A loose, inserted, continuous steel cooling ring placed between the cylinder wall and head carried heat away from the injector nozzle and scavenge valve seat areas. The ring contained drilled water passages.

e) Uncooled pockets in the piston were eliminated.

f) The cylinder and upper head were of cast iron; the lower head and piston were of three percent nickel cast steel.

Tests began February 1914 on the six cylinder engine using the last cylinder design. (Fig. 16) Cylinder problems unrelated to thermal loads delayed full power runs until September. The engine then produced 10,000 hp (7,457 kW) at 130 rpm. Work greatly slowed down then because of the war's starting.

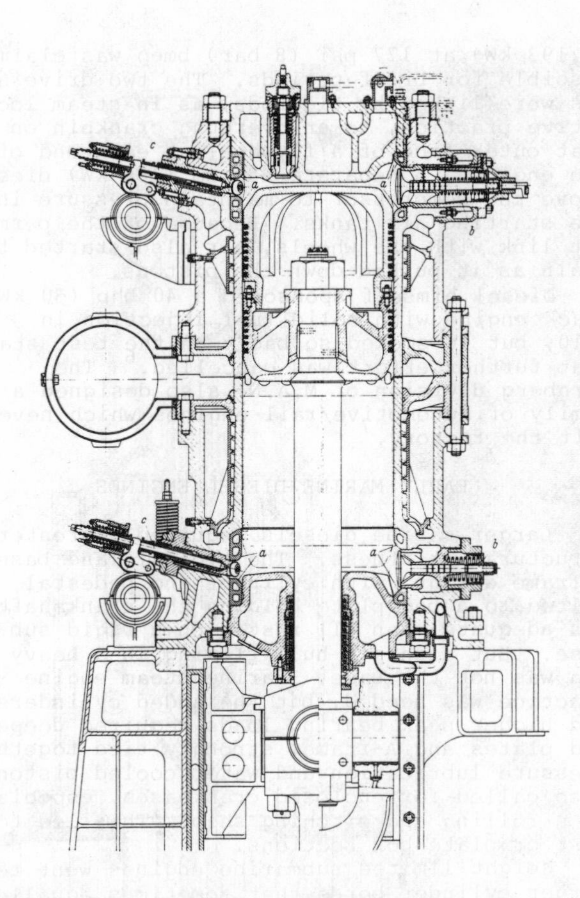

Fig. 14. M.A.N. 850 x 1,050 mm diesel, 1913. Seventh and last piston/cylinder design.

A shortage of gas-oil fuel caused further delays until changes could be made to use coal tar oil. Tests on the one cylinder engine with this fuel took place in April 1915. A five day run was also made at 2,030 hp (1,514 kW).

Six cylinder engine tests using a a tar oil/kerosene fuel mix finally began in January

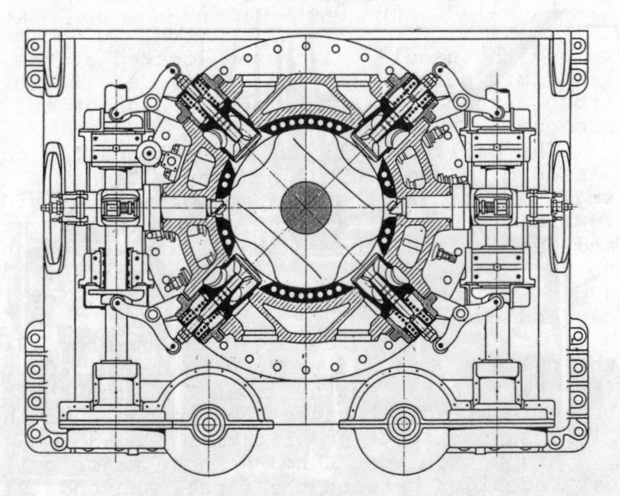

Fig. 15. M.A.N. 850 x 1,050 mm engine. Plan view through cylinder showing cooling ring and cam operated air valves. Last design, 1913.

1917. Two months later the engine passed the five day acceptance test at 10,800 hp (8,054 kW), or 90 percent power. Output was increased to 12,160 hp (9,068 kW) at 135 rpm for twelve hours. Indicated mep in both combustion chambers at this overload was about 9.6 bar. Specific fuel consumption was 243 g/hp hr (330 g/kW hr) with a mix of 214 g coal tar and 29 g kerosene. It is assumed that the indicated power of the injection air compressor was not deducted in calculating this bsfc. Tests on the reversing mechanism also proved satisfactory. The only noted problems were some broken piston rings. Other acceptance run data:[29]

Mean brake power...10,880 bhp (8,114 kW)
Mean ind. power.....14,500 ihp (10,812 kW)
Speed...............130 rpm
Bsfc -- g/bhp/hr*..228 (310 g/kW/hr)
Scav. pres. (abs)..1.59 kg/cm^2 (1.56 bar)
Injection air:
Compressor......823 ihp (613 kW)
Pressure (abs.).71 at (69.6 bar)
Water pump, etc.....300 ihp (224 kW)
Brake mech. eff#...69 %

* Based on a fuel heating value of 10,000 Cal/kg. Injection air compressor power not included.
\# All auxiliaries included.

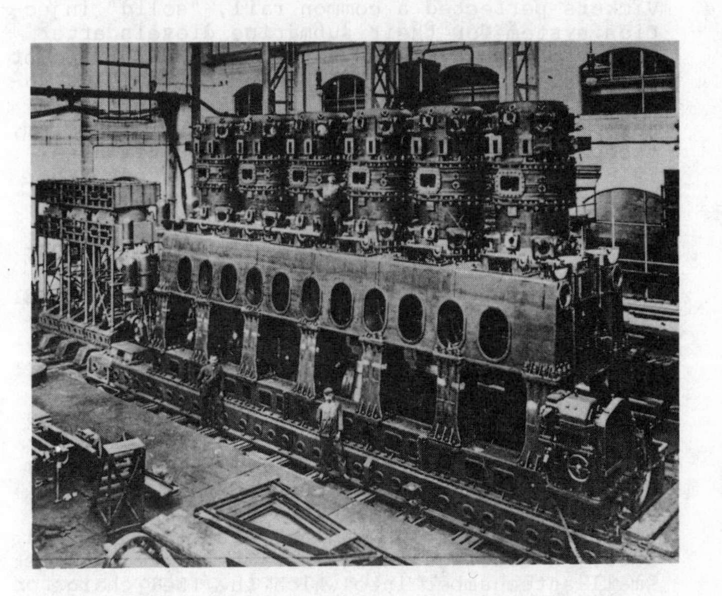

Fig. 16. M.A.N. 6 cylinder, double-acting, 850 x 1,050 mm engine during assembly, 1914. Note man standing by mid-engine cylinders.

The M.A.N. diesel was intended to replace the middle steam engine in the battleship Prinzregent Luitpold. The plan was for this to happen in 1913, but the disappointingly long development time, plus the war, changed everything. The ship could not be spared from service, so the engine never left Nürnberg and was scrapped after the armistice.

THE KRUPP ENGINE[30]

The Krupp Germaniawerft factory had a parallel program with M.A.N. Even though their design differed significantly Krupp, too, did

Fig. 17. Krupp 875 x 1,050 mm, 1 cylinder, double-acting, 2-stroke, engine. 1,490 kW/cyl.

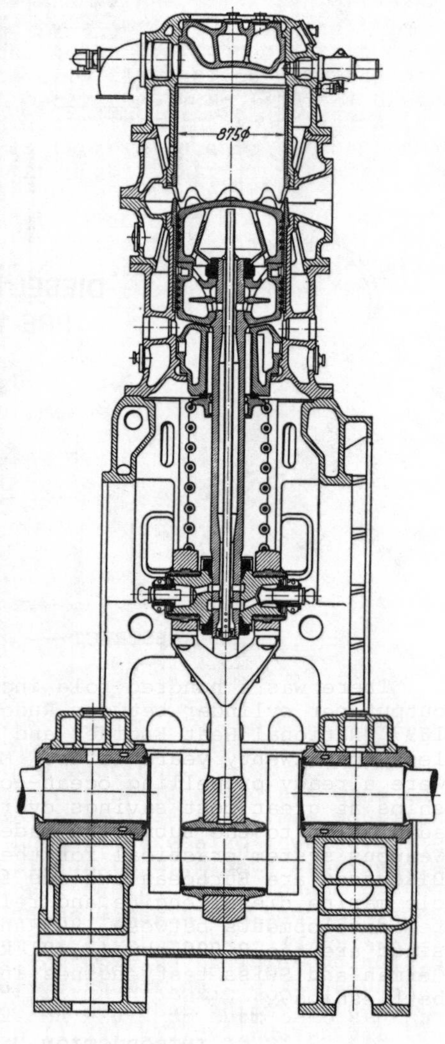

Fig. 18. Krupp 875 x 1,050 mm one cylinder, double-acting, 2-stroke test engine, 1911. First cylinder and piston design.

not pass the ninety percent power acceptance test until 1917.

Krupp delayed building a three cylinder version until a reasonable success first had been achieved on a single cylinder prototype. This engine became the workhorse. (Fig. 17)

The initial design of the 7.9 m high, double-acting 2-stroke engine included uniflow scavenging by four air inlet valves and central exhaust ports. (Fig. 18) The piston rod was guided by a single side crosshead guide. Scavenge air pumps were main engine driven, but all other pumps were operated by auxiliaries.

Hydraulic actuation of the scavenge valves and the two fuel valves was unique. (Fig. 19) Cams on a remote shaft created a timed opening force going to the desired valve via piping. Peak hydraulic pressures were about 80 bar. The stem of the outward opening scavenge valve was directly acted upon by a slave piston. A rocker lever lifted the inward opening injector needle valve against its spring seating pressure. A slave piston opposite the pivoted end raised the needle valve stem located at the lever mid point. The design gave so little trouble that it was retained throughout the program.

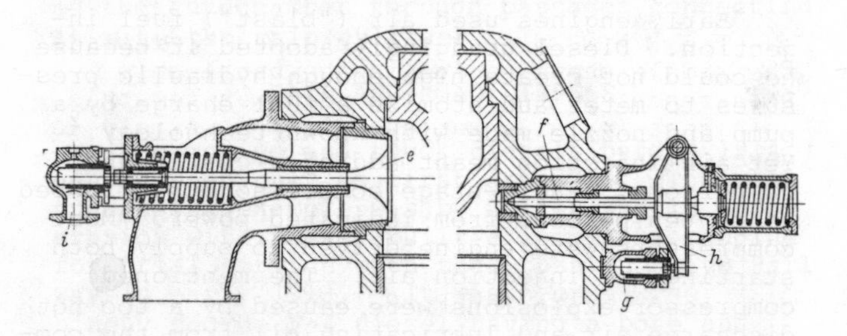

Fig. 19. Krupp 875 x 1,050 mm double-acting, 1911. Details of hydraulic scavenge air and fuel nozzle valve operation.

The built up cylinder assembly partly consisted of a two piece liner separated by a small end gap at the exhaust ports to provide for heat growth. Castings containing the air and fuel valves encased the liners and extended to the exhaust ports. They also retained the upper and lower heads which were merely cylinder end covers.

During tests which began in November 1911 the cylinder cracked under a load of 1,200 hp (895 kW). A few months later the crankshaft broke reportedly because the ground under the floor support settled slightly.[31]

Krupp then went through three major cylinder redesigns between the first test and 1917, with the last one only having been on the one cylinder engine. Their evolution included a combination of port and valve scavenging, next mainly port scavenging with small air valves retained near the head, and finally true loop scavenging with intake ports only. Much air flow work was done on the one cylinder engine to develop air and exhaust deflection profiles on the piston crowns. The inlet and exhaust ports were not angled.

The third design of 1913 as used on the six cylinder engine eliminated the poppet scavenge valves.

The last design added a thin shell contoured to maintain a small space between it and a wet liner through which coolant passed at a high velocity. It achieved the same effect as the spiral ribs in the M.A.N. engine. Reinforcements were also added inside the piston to prevent breaks at the rod connection. (Fig. 20)

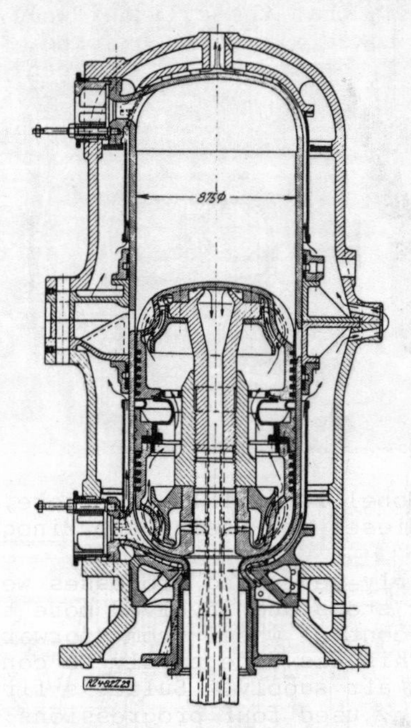

Fig. 20. Krupp 875 x 1,050 mm engine. Last design, 1913. Note piston cooling path.

In May 1913 the three cylinder engine was run under load. (Fig. 21) Problems, mainly in the cylinder area, fostered the above mentioned third design which was installed in January 1914. Official tests on the six cylinder engine did not begin until 1916, but the 90 percent power acceptance tests were completed with difficulty.[32] Performance data:

Mean brake power...10,600 bhp (7,904 kW)
Mean ind. power....13,640 ihp (10,170 kW)
Max. 1 hr power....12,060 bhp (8,990 kW)
Speed..............140 rpm
Bsfc -- g/bhp/hr*..215 (292 g/kW/hr)
(@ 7,904 kW)
Scavenge pump
Compressor......1,095 ihp (817 kW)
Press. (abs)....1.3 at (1.27 bar)
Injection air:
Compressor......2,000 ihp (1,490 kW)
Pressure (abs.).80 at (78.4 bar)
Mep @ 12,060 hp
Upper cyl.......7.25 at (7.11 bar)
Lower cyl.......5.50 at (5.39)
Water pump, etc....300 ihp (224 kW)
Brake mech. eff....63 % (incl. auxiliaries)

* Inj. air compressor power not included.

Note the difference in mean indicated pressure between the upper and lower cylinder sides. With the last design (not used on the full engine) the scavenge pressure dropped from 1.27 bar to 1.1 bar. This would have reduced the pump power and improved scavenging in the lower combustion chamber.

The engine was scrapped in 1919 as required by the Versailles Treaty.

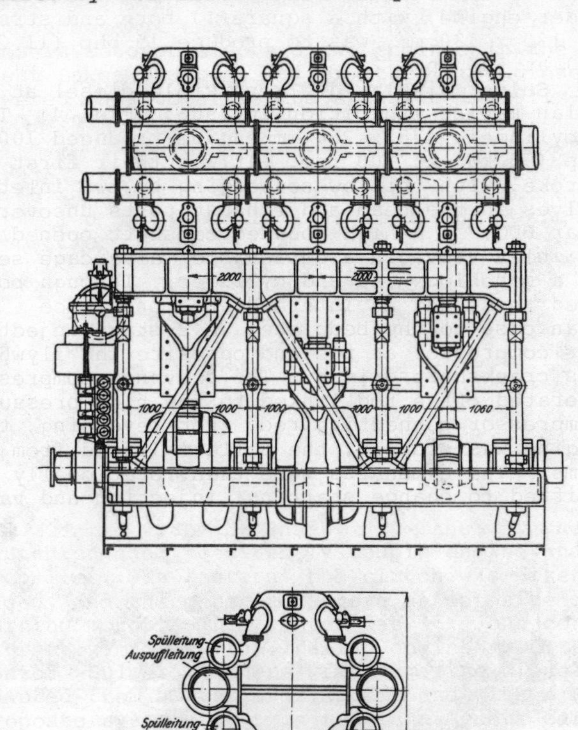

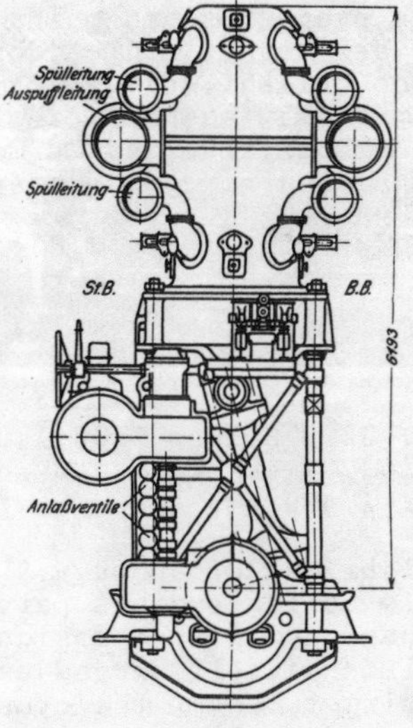

Fig. 21. Krupp 875 x 1,050 mm 3 cylinder, double-acting, 2-stroke engine, 1913.

THE SULZER S100 ENGINE[33]

Sulzer's S100 diesel, with its 1,000 x 1,100 mm bore and stroke, had a cylinder displacement over seven times larger than that of their largest marine engines entering service. (Fig. 22) In comparison, the Monte Penedo whose maiden voyage was in 1912, had two Sulzer 4 cylinder, 2-stroke engines of 850 bhp (634 kW) each with a bore and stroke of 470 x 680 mm. While the S100 might be viewed simply as a scaling up of production models, its great size increase alone presented problems requiring time to solve. The experience gained from this engine proved in valuable when Sulzer offered a new line of larger diesels after the war.

Being single-acting, its cylinder dimensions had to be much larger for the design power, and a multi-cylinder version would no doubt have been heavier than its rivals to the north, However, thermal loading and sealing problems were eased.

Fig. 22. Fig. 22. Sulzer S100 1,000 x 1,100 mm single-acting, 2-stroke, test engine, 1913. Design rating: 1,490 kW.

An overall height of about 6.9 m was less than the double-acting engines partly due to a compact crosshead and piston design. An extension of the piston itself served as the piston rod. Telescoping tubes on the crosshead guides and piston carried cooling water to and from the piston. (Fig. 23) One reported problem was with the crosshead guide bearings which had to be increased in area.[34]

The Sulzer "cross port" scavenging continued with its two rows of intake ports as introduced on the <u>Romanga</u> and described earlier. The upper intake ports received air which passed a cam operated valve in the scavenge air supply passage. The cylinder itself consisted of a flanged wet liner and an exterior sleeve to form water, air and exhaust passages.

There was only one fuel injection valve and a starting air valve in the cylinder head, a simple dome over the piston. Because of the great bore diameter the head depth (approx. 0.8 m) had to be enough to structurally carry the combustion pressure loads. Tie rods extending from the head to the bottom of the bed plate carried the axial loads and held all together.

Originally the scavenge air compressor and the three stage air compressor for injection were driven by the engine, but later they were auxiliary powered.

Design of the engine started in 1910 and testing began in 1913. In addition to the crosshead bearing problem there were also difficulties with the single injection nozzle. Once early in the tests the fuel was injected too soon to ignite for several revolutions. When ignition did occur the accumulation of fuel created excessive pressure and a longer burning such that the cylinder wall between the upper and lower air ports cracked. A "patch" fitted over the cracks put the engine back in operation.[35]

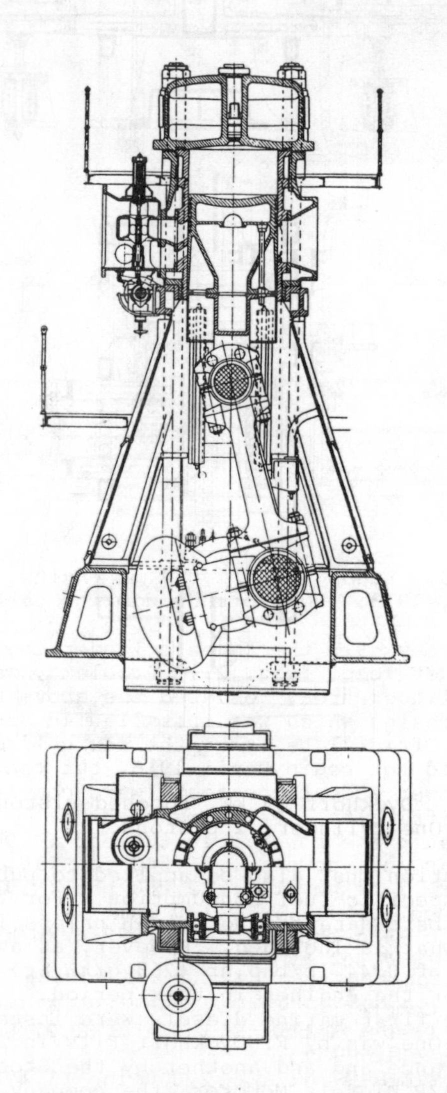

Fig. 23. Sulzer S100 1,000 x 1,100 mm engine, 1913. Cam and rocker lever for fuel valve and starting valve actuation not shown.

Prof. Stodola conducted independent tests in November 1914 and January 1915 which showed the engine had a consistently low bsfc: for example, 198.1 g/bhp hr (269.3 g/kW hr) when

Fig. 24. M.A.N. 10 cylinder, 300 x 300 mm 4-stroke, submarine diesel, 1917. 2,237 kW at 390 rpm.

delivering 1,986 bhp (1,481 kW) at 149 rpm. The bmep at this rating was 6.97 kg/cm^2 (6.83 bar), not including the compressor engine. Total indicated horsepower of the main and auxiliary engines was 2,812 ihp (2097 kW). This gave a mechanical efficiency of 71 percent. The fuel was a diesel oil (Galician) with a specific gravity of 0.871. Disposition of the engine is not known.

CONCLUSION

Marine diesel engines made great progress in performance and reliability after their introduction in 1903. During 1913 more than 22 diesel powered ships of from 2,000 to 10,000 tons were being built.[36] Most were twin screw, with the largest having 1,750 bhp (1,305 kW) per shaft. Many of the major European builders were running concurrent surface and submarine diesel programs. Two and 4-stroke engines began competing, with the 2-stroke starting to gain in favor for larger outputs. Only the Vickers had a fuel injection system not requiring a high pressure air compressor. Crosshead engines were standard for surface ship and the trunk piston design was mandatory in submarines. The very large diesels conceived for battleships had pushed the state of the art in designing for high thermal and combustion pressure induced stresses. Not addressed are the submarine diesels produced in the latter part of World War I that exhibited a high degree of sophistication. A notable example was the ten cylinder, 3,000 hp (2,237 kW) M.A.N. diesels that became the design target for navies which later gained possession of these engines. (Fig. 24)

Another of Rudolf Diesel's prophesies made in 1892 came to pass with the growing acceptance of marine diesels as a workhorse engine.

REFERENCES

1. Supino, G, Land & Marine Diesel Engines, Griffin, London, 1918, pp. 24-26.
2. Diesel, R., Die Entstehung des Dieselmotors, Springer, Berlin, 1913, p. 24.
3. Supino, op. cit., p. 132.
4. Hausfelder, L., Die Kompressorlose Dieselmaschine, Krayn, Berlin, 1928, pp. 287-298. Also, Vickers literature of the period kindly given to the author.
5. "The Sulzer One Thousand Horse-Power Diesel-Engined Locomotive", Internal Combustion Engineering, Sept. 17, 1913, Sulzer 1913 reprint.
6. SAFIR (Schweizerische Automobil-Fabrik in Rheinneck) records kindly supplied by George Aue of Winterthur. This Swiss company built small engines to Diesel's design in the 1910-13 period.
7. Schnauffer, K, "Die Dieselmotorenentwicklung in Werk Nürnberg der M.A.N. 1897-1918", mss., 1956, pp. 6-12.
8. Supino, op. cit, p. 18.
9. Diesel, R., "The Diesel Oil-Engine", Engineering, pp. 398, 402.
10. Flasche, H., "Die ersten Dieselmaschinen-Schiffe der Welt", mss, Kiel, 1947. Hans Flasche, Rudolf Diesel's brother-in-law, worked at Nobel's beginning in 1899.
11. Ostertag, P., "Der Lastdampfer 'Venoge' auf dem Genfersee," Schweizerischen Bauzeitung, Vol. 48, No. 13, 1907, Sulzer reprint.
12. Technical information in M.A.N. archives.
13. Drawings in Sulzer Archives.
14. Flasche, H., "Der Dieselmaschinenbau in Russland von Anbeginn 1899 an bis etwa Jahresschluss 1918", mss, pp. 16-17.
15. Supino, op. cit., p. 184.

16. Technical and operational information in Sulzer archives.
17. Burmeister & Wain's Marine Diesel Engines, B&W, Copenhagen, 1920. Also The Gas Engine, April 1912, pp. 175-182.
18. Regenbogen, C., "Der Dieselmotorenbau auf der Germaniawerft", Schiffbautechnischen Gesellschaft, 1913, reprint, pp. 5-6.
19. Sass, F., Geschichte des Deutschen Verbrennungsmotorenbaues von 1860-1918, Springer, Berlin, 1962, pp. 559-576.
20. König, P., Voyage of the Deutschland, Hearst, New York, 1916, 247 p., translation. (König was the skipper of the sub.)
21. "Motor Tankship Wotan", The Gas Engine, Jan. 1914, p. 39.
22. Sass, op. cit., p. 310.
23. "An 800 h.p. Marine Engine", The Gas Engine, July 1913, p. 360. Also, Milton, J.T., "Present Position of Diesel Engines for Marine Purposes", Trans., Inst. of Naval Architects, 1914, p. 88.
24. Blunck, R., Hugo Junkers: der Mensch und das Werk, Limpert, Berlin, 1942, p. 72.
25. Magdeburger, E., "Diesel Engines in Submarines", Trans., Am. Soc. of Naval Engineers, Vol. 37, No. 3, Aug. 1925, p. 582.
26. Ibid., p. 580.
27. Sass, op. cit., p. 551.
28. The major source for the M.A.N. engine is K. Schnauffer, "Die Dieselmotorenentwicklung in Werk Nürnberg der M.A.N. 1897-1918," 1956, mss, pp. 41-55. He drew from a 281 page test report kept by the German Admiralty. Schnauffer's work was incorporated in Sass, op. cit.
29. Seiliger, M., Die Hochleistungs-Dieselmotoren, Springer, Berlin, 1926, p. 154. The first column are the original units.
30. Krupp engine sources: Seiliger, ibid., pp 156- 58; Sass, op. cit. pp. 576-80; and Krupp Marine Diesel Engines, Krupp Germaniawerft, ca. 1930, p. 7.
31. Nägel, The Diesel Engine of To-day", Diesel Engines, V.D.I., 1923, p. 17.
32. Seiliger, M., op. cit., p. 157.
33. Sulzer engine sources include: Performance data in Sulzer Archives; "The Sulzer "Two-Stroke Marine Diesel Engine", Sulzer Technical Review, 1947, No. 2, pp. 19-21; and Seiliger, M., op. cit., p. 158-60.
34. "The Sulzer 2,000 bhp Single Cylinder Motor", The Gas Engine, Aug. 1913, p.446;
35. Hawkes, C., "Note on the present position of the various experimental oil engines of large power per cylinder", Royal Commission on Fuels & Engines, London, 1913, p. 11.
36. Milton, J., "Summary -- Engines and Tank Vessels", Royal Commission on Fuels & Engines, London, 1913, p. 43.

DRAWING AND PHOTO CREDITS

Chalkey, A. Diesel Engines for Land and Marine Work, 4th ed., 1919: 11, 17.
Lehmann, J., Rudolf Diesel and Burmeister & Wain, 1938: 8, 9.
Maag, J., Dieselmaschinen, 1928: 7, 12, 18, 20, 21.
M.A.N. Archives, Augsburg: 2, 16, 24.
Nägel, A., "The Diesel Engine of Today", Diesel Engines, (V.D.I.): 15.
Sass, F., Geschichte des Deutschen Verbrennungsmotorenbaues von 1860-1918, 1962: 10, 13, 14, 19.
Seiliger, M, Die Hochleistungs-Dieselmotoren, 1926: 23.
Sulzer Archives, Winterthur: 3, 6, 22.
Trans., Inst. Engineers & Shipbuilders, (Scotland), 1923: 4.
Trans., V.D.I., (Berlin), 1903: 1.
Wells, J. & Wallis-Taylor, A., The Diesel or Slow Combustion Oil Engine, 1914: 5.

THE AUTOMOTIVE SPARK-IGNITION ENGINE – AN HISTORICAL PERSPECTIVE

C. A. Amann
General Motors Research Laboratories
Warren, Michigan

ABSTRACT

The foundation for the four-stroke-cycle reciprocating gasoline engine that dominates the automotive field today was laid by the early 1900s. During the first half of this century there was a major focus on avoidance of combustion knock, which blocked the route to increased efficiency and specific power. Research led to the discovery of tetraethyl lead, pointed the way to improved gasolines, and influenced combustion-chamber design. Starting in the 1950s, emissions became a major concern. Mixture preparation and spark timing, once mainly manually controlled, had become automatically regulated, using mechanics, hydraulics, or pneumatics. Since control of emissions has necessitated integration of the catalytic converter and its on-board computer into the engine system, such engine-control functions are now increasingly done electronically. Modern fuel economy, specific power, exhaust emissions, and driver convenience would astound the early automotive pioneers.

INTRODUCTION

The automobile had its start in Europe late in the nineteenth century. It was not invented by any one individual, but represented a clever amalgamation of related developments. Two of these were construction of an engine sufficiently compact to be mounted in a vehicle, and ready availability of a liquid fuel for that engine.

Around the turn of the century, the automobile was eagerly anticipated by many Americans who had never seen one. If for no other reason, a way to ease the solid-waste-disposal problem with horses (22 lb/horse-day), along with the associated flies and odor, was a welcome prospect. However, as the automobile emerged, it often faced repressive laws intended to preserve the status quo. In England in 1865, when the steam carriage began to appear in significant numbers on public roads, Parliament had decreed that a self-propelled highway vehicle had to be preceded by a man bearing a red flag and walking 60 yards ahead at a speed of no more than 4 mi/h [1]. Later in the U.S., speed limits of under 10 mi/h were often set for the fledgling automobile. Police sometimes stretched a cable across the road to restrain speeders, and soon private citizens were following suit. This led inventors to propose scissors mechanisms for mounting on the fronts of cars. Anti-auto zealots threw rocks at motorists, or spread broken glass and nails on the road. One group proposed that automobiles traveling at night on rural roads fire warning rockets and Roman candles [2]. With such an inauspicious beginning, who would have foreseen that within less than a century, there would be an automobile in the U.S. for every two people.

Today well over 90% of those cars are powered by four-stroke-cycle spark-ignition engines, but that was not always the case. Of the American cars built in 1900, 1681 were powered by steam, 1575 were electrics, and only 936 used spark-ignition engines [3]. As the spark-ignition engine was improved, however, the disadvantages of the competitive propulsion systems became unacceptable.

For example, in those early days it took 20 to 45 minutes to raise pressure for the steam car, while the internal combustion engine could be started essentially instantaneously. The early steam car lacked a condenser, so frequent stops for makeup water were required. The electric car started promptly, like the internal combustion engine, and it was appreciated for its comparative silence. However, it was expensive, slow, and suffered from limited range. Moreover, storage batteries could not be recharged as quickly as a fuel tank could be refilled. In those early days, the operating radius of a car powered by a spark-ignition engine (for one tank of fuel) was nearly triple that of an electric (before requiring battery recharging) and four times that of a steam car (before requiring water) [4].

The spark-ignition engine had one outstanding disadvantage relative to steam and electricity, however. The repetitive explosive noise of the exhaust was likened to that of a machine gun. It frightened the then ubiquitous horse, sometimes inciting it to rear. Early introduction of the exhaust muffler made the engine somewhat more compatible.

Although the internal-combustion engine could be started promptly under most circumstances, the standard method -- a hand crank -- was understandably unpopular. With improper spark timing, the engine backfired, all too often breaking the arm of the person turning the crank.

An accident of that nature fostered introduction of the electric starter in 1912. A friend of Henry Leland, head of Cadillac, suffered a broken jaw from an engine backfire while cranking a car and died from ensuing complications. C. F. Kettering's offer to develop an electric starter for Leland shortly thereafter was welcomed enthusiastically. Kettering's success stemmed in part from his recognition that the starting motor did not have to be as big as a motor of the same capacity designed for continuous operation because of the intermittency of the starting requirement [5]. Eliminating the need for hand cranking paved the way for the woman driver.

THE FOUR-STROKE CYCLE

The four-stroke cycle followed by most automotive engines today is usually credited to a German traveling salesman, Nicolaus Otto, who was intrigued by the possibilities of the internal-combustion engine [6]. Although he demonstrated the first working four-stroke engine in 1876, the principles of the four-stroke cycle were originally delineated by a Frenchman, Beau de Rochas, in 1862.

Otto's early experiments were with the Lenoir two-stroke engine of 1860, which inducted a mixture of coal gas and air during the first half of each piston downstroke, then spark ignited the mixture without benefit of compression, using the rest of the downstroke for expansion and the return stroke for exhausting the burned gases from the cylinder. From there Otto progressed to an atmospheric engine, in which explosive combustion drove a free piston vertically upward, after which the cooling of the hot gases caused the piston to fall and turn the output shaft through a free-wheeling clutch that was disengaged during the upstroke of the piston. This engine was produced and marketed in quantity.

Otto's subsequent four-stroke engine consisted of a single horizontal cylinder with a bore and stroke of 161 x 300 mm (6.3 x 11.8 in.) and a compression ratio of only 2.5. It used slide valves, reminiscent of the steam engine, and employed a flame for ignition. Run on illuminating gas, it developed 2.2 kW (3 hp) at a speed of 180 r/min. Its thermal efficiency of 14% was two to three times that of a comparable steam engine [6].

This four-stroke engine was much quieter than Otto's atmospheric engine and became known as the "Silent Otto." Otto erroneously attributed the comparative silence of his engine to charge stratification, believing he had layered a fuel-rich mixture adjacent to the cylinder head for easy ignition and residual gas against the piston crown, with fresh air separating the two layers. Sir Dugald Clerk, who was himself later responsible for a successful two-stroke engine, overturned that belief in 1885 through experiments with smoke in a transparent cylinder that revealed the turmoil produced in the cylinder gas during induction of the fresh charge [7].

Although Otto's clumsy four-stroke engine was clearly unsuited for automotive use, it laid the foundation for those that followed. A cross section through the noteworthy 1899 single-cylinder automotive engine of Gottlieb Daimler is shown in Fig. 1 [8]. The intake and exhaust valves oppose one another in an antechamber connected to the space above the piston by a passageway. The self-actuated intake valve at the top of the antechamber is drawn open against a compression spring by the vacuum created by the descent of the piston on the intake stroke. The exhaust valve at the bottom of the antechamber is cam actuated through a push rod. The lower end of the pushrod is wedge-shaped, with the tip fitting into a notch in the side of the cam follower. When the engine overspeeds, a governor linkage moves the pushrod to the right, out of the notch, and the exhaust valve remains closed until the engine speed drops below its governed maximum.

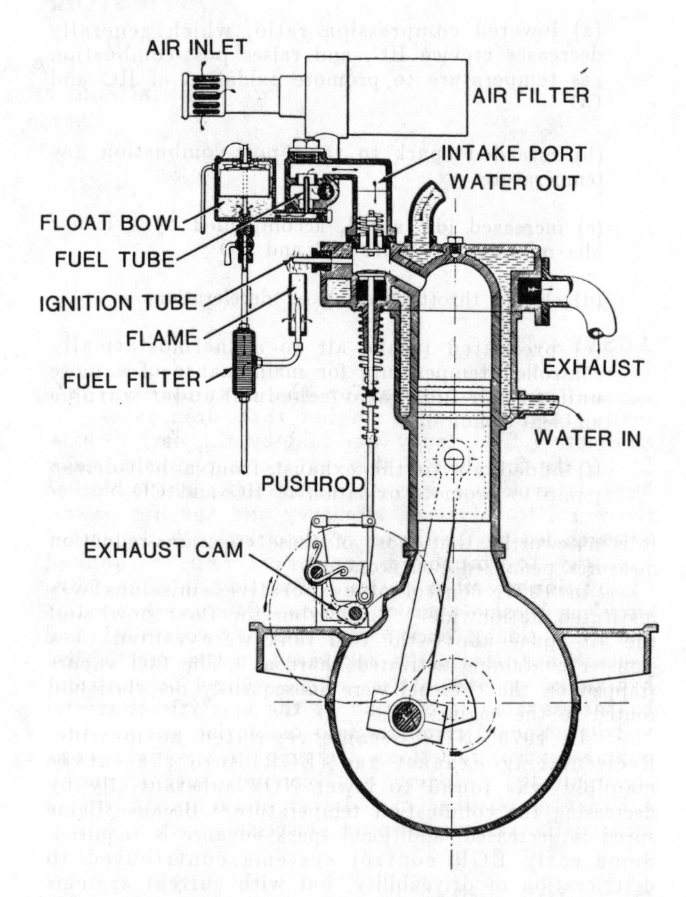

Fig. 1 Daimler engine of 1899 [8]

Air enters the engine at the upper left through adjustable slots. After going through the filter at the upper right, the air passes through a circular passage normal to the plane of the drawing and enters the intake port by flowing upward through an annulus surrounding a vertical fuel tube. The tube is filled from the bottom with fuel from the float bowl to its left. The depression created by the air flowing through the annulus draws fuel from the tube to mix with the air as it heads for the intake valve. This arrangement is similar in principle to that of the traditional carburetor. Other carburetors of that day created the fuel-air mixture by bubbling the air through a fuel reservoir.

The mixture is ignited near top dead center by means of an ignition tube seen protruding from the left of the antechamber between the two valves. This tube is filled with air-fuel mixture as the piston ascends on the compression stroke. The tube is heated externally by a continuous flame. The lack of control of combustion timing with this arrangement is striking. Flame-tube ignition was also considered a fire hazard.

These objections were overcome by the electric ignition systems of that day, which employed a spark plug surprisingly similar to those in use today. Both battery and magneto systems were in use.

The battery was either of the dry-cell type, requiring replacement after 300 to 400 hours, or of the storage-cell type, which necessitated recharging after 25 to 30 hours [8] because there was no engine-driven generator for on-board recharging at that time. An induction coil was used to provide the ignition pulse to the spark plug. Breaker points were sometimes used to interrupt the current in the primary circuit, as later became standard practice for many years, but in those early days this system was considered unreliable because it normally produced only a single significant spark on each engine cycle.

As an alternative, Karl Benz incorporated a vibrator in the system to make and break the primary circuit several times during the time interval when that circuit was connected to the battery by a rotating switch. The timing of the closing of that switch could be varied by the operator. The result was a multi-strike, controlled-timing ignition system.

Robert Bosch was instrumental in developing an early magneto-ignition system. This low-tension system replaced the battery with an engine-driven electric machine and also offered flexibility in spark timing. Daimler switched from flame-tube ignition to the magneto in 1899 [9].

What emerges from this brief snapshot of the automotive engine at the turn of the century is that the key elements of the reciprocating spark-ignition automotive engine as we know it today were all worked out in primitive form in those early days. What has happened since that time is a succession of refinements that has led to thermal efficiencies and power levels that would have astounded those early pioneers, concurrent with a cleaning of the exhaust gases that was not even on their minds.

FUELS AND COMBUSTION KNOCK

An historical review of the spark-ignition engine would be incomplete without considering the role played by early fuels development. The gaseous fuel used in Otto's original four-stroke engine was, of course, unsuited to vehicle application. Early petroleum-based fuels often caused a mysterious combustion knock, even at the low compression ratios of that day. Thomas Midgley, who was instrumental in discovering the anti-knock agent, tetraethyl lead, once described knock as "the noise that could be heard across a ten-acre lot" [7].

Early researchers sought an understanding of knock through measurement of instantaneous cylinder pressure. For example, Sir Harry Ricardo told how as a student at Cambridge University in the early 1900s, he joined Professor Bertram Hopkinson in applying an optical pressure indicator common in that pre-electronics age that relied on reflection of a light beam by a mirror that was deflected by cylinder pressure [10]. Unfortunately, every time the engine knocked, the mirror either shattered or was thrown from its frame. About 1913 Ricardo, continuing the experiments in his garden workshop, successfully showed that cylinder pressure rose perfectly normally until knock occurred, whereupon it "shot up abruptly." Ricardo's garden experiments seemed to resolve an existing controversy over whether knock started before or after the spark.

A few years later in a converted Ohio tobacco warehouse, C. F. Kettering and his team also installed an optical indicator on an engine [11]. T. A. Boyd, a member of that team, told how photographic paper was wrapped around a tomato can and secured with rubber bands. With the can supported on its axis between shingle-nail pivots, Kettering hand-spun the can while Midgley operated the shutter of the indicator. They were rewarded with a pressure-time trace showing clearly the severe oscillations in cylinder pressure responsible for the noise of knock.

In attempting to gain a deeper understanding of knock, many investigators turned to observations of flame travel through small windows in the cylinder head. Finally in 1936, Withrow and Rassweiler acquired the first full-field movies of knocking combustion through a quartz cylinder head [12]. The pictures confirmed that knock resulted from autoignition at some point ahead of the flame front, sometimes more than one, followed by almost instantaneous inflammation of the end gas.

Turning back to the days when knock was still a mystery, Ricardo told how "the discriminating motorist never bought a drop of gasoline without first dipping his finger in the can and blowing on it to test its volatility" [10]. This was indeed a crude way to separate knock-prone kerosene from the preferred more volatile fuel, gasoline.

This preoccupation with volatility explains the line of logic that led to discovery of the antiknock qualities imparted by tetraethyl lead [11]. The Kettering team noted that the trailing arbutus, a wild flower with red-backed leaves, was an early spring bloomer. They suspected this was because red was a superior absorber of radiation from the sun. They reasoned that if kerosene were dyed red, the unburned kerosene-air mixture ahead of the flame front might absorb heat from the flame more readily, vaporizing the kerosene better and causing it to resist knock like gasoline. To test this hypothesis, a handy bottle of iodine was removed from the laboratory shelf and its contents used to color kerosene. When the mixture was run in the engine, knock was greatly decreased. That success was short lived, however, because when ordinary red dye was subsequently tried, it had absolutely no effect on knock. However, the success with iodine pointed to the potential of a chemical additive as a knock suppressor.

Aniline-type compounds proved more effective than iodine but produced a smelly exhaust. Tellurium was even better but smelled like "a mixture of garlic and onions," with the odor only worsened by trying to wash it from hands and clothing. Finally, using the periodic table, a systematic sorting of possible compounds was undertaken that uncovered the antiknock properties of tetraethyl lead (TEL). The damage TEL did to exhaust valves and spark plugs was alleviated by adding organic compounds of bromine and/or chlorine.

In the course of this research it was noticed that the chemical structure of the fuel is important in determining its propensity to knock [11]. The anesthetic, ethyl ether, was particularly knock-prone, while butyl alcohol was quite knock-resistant, yet both share the chemical formula $C_2H_{10}O$. Later many researchers became involved in systematically examining the knocking tendencies of a large variety of hydrocarbon fuels, leading to a general understanding of the contribution of molecular structure [13].

As the automobile population multiplied, a formal procedure for ranking the knock resistance of the various hydrocarbon fuels became a necessity. This topic occupied a subcommittee of the Cooperative Fuel Research (CFR) Committee starting in the late 1920s. Graham Edgar of the Ethyl Gasoline Corporation had prepared samples of a variety of pure hydrocarbons, and these were evaluated in a single-cylinder variable-compression-ratio engine by John Campbell at General Motors, which had built the nation's only engine of that type at the time. Campbell compared evaluation of fuels in this engine to "having a 100-in. telescope to explore the heavens -- a field wide open for exploration" [14]. In 1929, T.A. Boyd urged the CFR subcommittee to adopt such a variable-compression-ratio engine for rating gasolines. There was considerable resistance to the proposal because the engine was viewed as too complex, but Harry Horning, president of the Waukesha Motor Company, promised to build such an engine for trial. This CFR engine was displayed at a meeting of the American Petroleum Institute in November 1931 and ultimately provided a basis for rating fuel knock characteristics. About 5000 of these CFR engines have been built since then, and in 1980 the ASME designated the engine as its 49th mechanical engineering landmark.

While the CFR engine provided the apparatus whereby the knock resistance of a fuel could be assessed, it remained to quantify that quality. The octane rating scale was developed, based on the the primary reference fuels normal heptane, assigned an octane rating of zero, and 2-2-4 trimethyl pentane, an isomer of octane assigned a rating of 100. Gasolines with knock tendencies intermediate between these two primary reference fuels are generally characterized by a Research Octane Number (RON) and/or a Motor Octane Number (MON). The former, which is measured under different operating conditions in the CFR engine than the latter, is generally the higher of the two.

MECHANICAL OCTANE NUMBERS

Evolutionary advances in petroleum refining and the availability of TEL for many years led to increasing fuel octane rating, with a corresponding increase in compression ratio for higher engine thermal efficiency and specific power. These engine improvements were not attributable solely to the fuel, however. Engine designers also learned how to configure combustion chambers for lower octane requirement, i.e., how to design mechanical octane numbers into them.

For example, the combustion chamber of an early side-valve engine (circa 1920) having both valves on the same side of the cylinder is illustrated in Fig. 2. As late as 1928, nearly 2/3 of U.S. passenger-car engines employed this valve arrangement. One used the inlet-over-exhaust configuration typified by Fig. 1, two used sleeve valves, and less than a quarter used the overhead-valve arrangement found universally today [15].

In 1923 Ricardo patented the "Turbulent Head" side-valve engine that is shown in Fig. 3. The reduced knock tendency of this chamber shape, relative to that of Fig. 2, is generally attributed to (a) increased turbulence for faster flame speed through "squish" of the end gas as the piston nears the end of the compression stroke, (b) enhanced cooling of the end gas, and (c) better concentration of the mixture around the spark plug for a faster mass-burning rate.

Following World War II, C. F. Kettering compared an experimental 6-cylinder in-line engine with a compression ratio of 12.5 to the then-current production engine with a compression ratio of 6.4 [16]. His purpose was to illustrate the improvement possible if higher-octane gasoline could be made available. The side-valve configuration of the production engine was replaced with overhead valves in the experimental engine for increased structural rigidity to overcome anticipated problems with engine roughness and increased friction, for the peak cylinder pressure was almost doubled in the high-compression engine. At high compression ratios, the reduced clearance volume also limits space for operation of the valves in the side-valve geometry, and the reduced projected area of the head exposed to cylinder pressure eases the head-gasket sealing problem in the overhead-valve configuration. Kettering's high-compression engine delivered almost 50% more power per unit displacement and consistently showed an improvement of about 35% in fuel economy.

The oil industry responded to this challenge with increased-octane fuel, and the automotive industry responded with higher compression ratios. As an example of what could be achieved through combustion-chamber design, Caris et al. catalogued the octane requirements of a wide variety of shapes [17]. As illustrated in Fig. 4, they found that compacting the chamber around the spark plug generally produced beneficial effects. It is seen to the right on the octane-requirement yardstick of Fig. 4 that when the clearance volume is stretched out from the centrally mounted

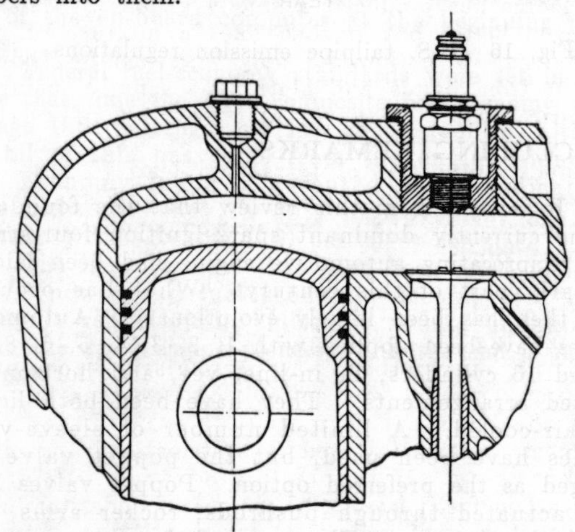

Fig. 2 Side-valve engine, circa 1920 [15]

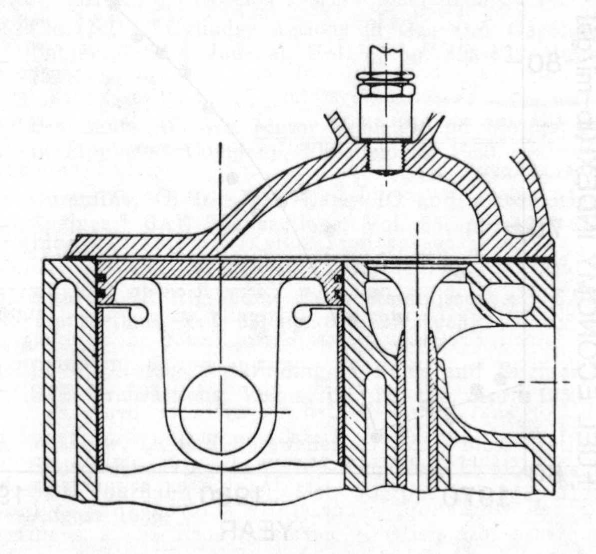

Fig. 3 Ricardo "Turbulent Head" [15]

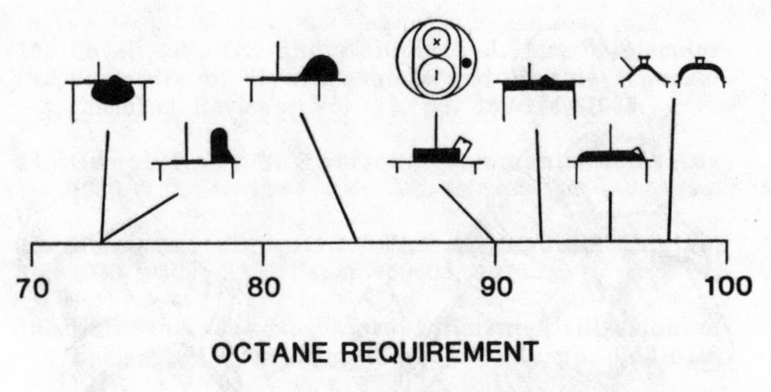

Fig. 4 Chamber shape and octane requirement [17]

spark plug, a 98-octane fuel is required to avoid knock. On the opposite end of the scale, when the chamber volume is concentrated near the spark plug, the octane requirement drops to 73.

The course of average compression ratio in U.S. cars is plotted in Fig. 5 from the time when TEL was introduced until when unleaded fuel was restored to the market. Termination of the upward trend in the late 1950s coincided with the onset of an abnormal combustion phenomenon termed "rumble" that was encountered in engines with a compression ratio of 10 or more. It occurred at full load and caused an objectionable noise of lower frequency than knock. Rumble was traced to the initiation of multiple flame fronts by incandescent combustion-chamber deposits [18].

Nationwide control of tailpipe emissions became effective in 1968 and was met with a hasty retreat in average compression ratio, as seen in Fig. 5. Unleaded 91 RON gasoline was made available in 1973 in anticipation of the introduction of catalytic converters, which cannot tolerate leaded fuel. By then, average compression ratio had retreated to its level of 17-18 years earlier. Since that time, the average U.S. compression ratio in 1988 has crept up to almost 9, giving an estimated one-third increase in indicated thermal efficiency over the level of the mid-1920s. The increase in average compression ratio of engines operating on unleaded regular fuel since 1973 has been primarily the result of increased mechanical octane number, although petroleum refiners have concurrently been able to boost the average chemical octane number of unleaded fuel as well.

To increase mechanical octane number, combustion-chamber designs have been refined, with emphasis being placed on faster burning of the charge so that the end gas has less time to experience the pre-reactions that trigger knock. Two different approaches are illustrated in Fig. 6 and 7. The use of intake swirl is represented by Fig. 6. In a proper design the large-scale swirling motion breaks down into finer-scale turbulence during the compression stroke, yielding a faster flame speed. In Fig. 7, the two pairs of inclined valves facilitate a central spark-plug location for minimum flame-travel distance. Speculation also exists that the inclined intake valves foster a tumbling motion capable of enhancing turbulence during compression.

Electronics has been enlisted to increase mechanical octane number. Electronic knock control is used on many engines today. This approach involves detection of incipient knock by an engine-mounted sensor, which then signals the on-board computer to retard spark timing enough to avoid it. Because knock is normally a problem only at high engine loads, this allows operation with retarded spark timing at a higher compression ratio than would otherwise be tolerable, while capitalizing fully on the higher compression ratio at lighter loads with the spark set at best-economy timing.

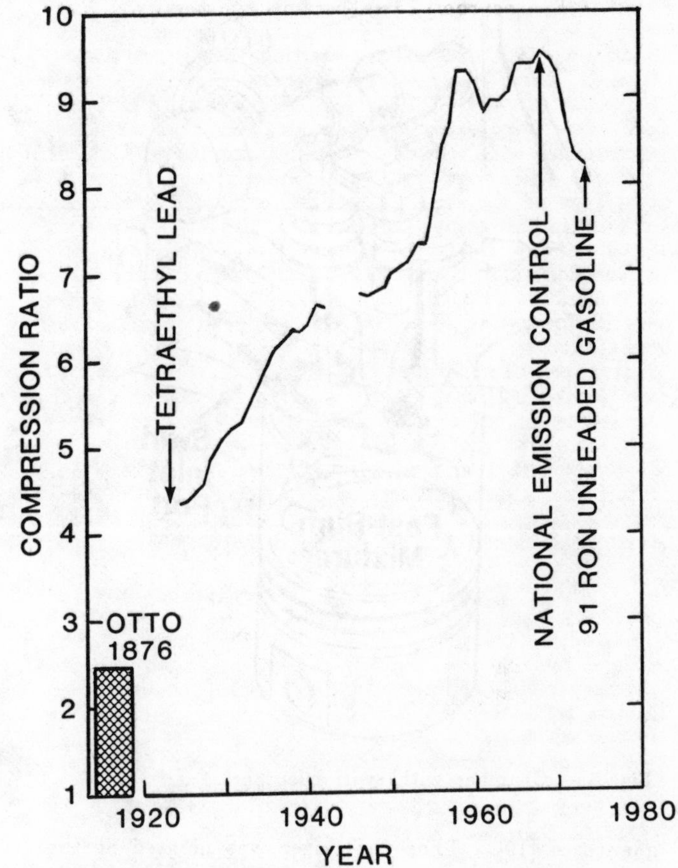

Fig. 5 Average U.S. compression ratio, historical trend

As for chemical octane number, petroleum refiners are now able to offer some unleaded premium gasoline with an (R + M)/2 rating (where R and M represent the Research and Motor octane numbers, respectively) of 91 and above. This compares to a typical (R + M)/2 for regular unleaded gasoline of 87. Most engines are still designed to operate on unleaded regular, however.

DAWN OF THE EMISSIONS ERA

Thanks to its particular topography and its abundant sunshine, the Los Angeles basin has been noted for abnormal air quality since the time when it was inhabited exclusively by the native American Indian. By the early 1950s, industrialization and population growth had reached the point where occurrences of "smog," most immediately evidenced by eye irritation, were sufficiently frequent to arouse the citizenry. Ozone was linked to the problem. At the California Institute of Technology, Professor A. J. Haagen-Smit had been able to show in his laboratory that ozone could be produced by exposing selected hydrocarbon vapors and nitrogen dioxide mixed in proper proportions to sunlight. Both of these chemicals are found in automotive exhaust gases in minute

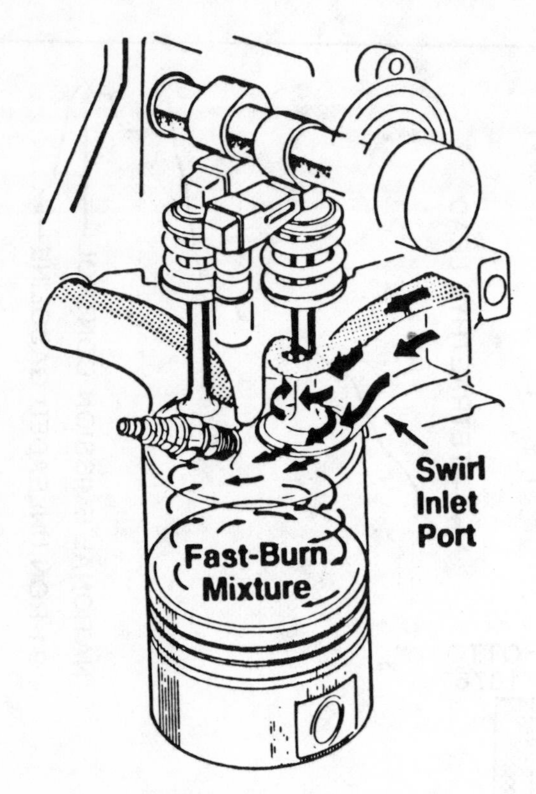

Fig. 6 Chamber with swirl inlet port

quantities [19]. Thus the finger was pointed at the automobile, and Los Angeles wanted an immediate solution.

In 1954 a group of experts from the automotive industry went to California to learn of Haagen-Smit's findings first-hand [20]. They were impressed with the quality of the work but equally surprised by the possibility that the automobile might actually be a significant contributor to the problem. At a closing meeting with a Citizen's Committee, the industry group burned a pot containing a mixture of gasoline, oil, vegetable clippings, and a few pieces of hamburger to simulate the hundreds of incinerators then operating in Los Angeles. Then they burned a gasoline torch to simulate the automobile. This comparison provoked an angry response from members of the Citizen's Committee, led to an eventual ban on incinerators, but did not resolve the issue of automobile exhaust.

There followed a period of frenzied activity in Detroit aimed at understanding the sources of the minute and previously largely ignored sources of unburned hydrocarbons (HC) and oxides of nitrogen (NOx) in the exhaust gas so that something could be done about them. Incomplete combustion at high manifold vacuums was found to cause high exhaust HC concentrations during deceleration [21], the driving mode of primary suspicion according to early tests. Wayne Daniel photographed the quench layer on the wall of a combustion chamber in an operating engine and suggested that as a source of HC during normal operation [22]. Because of the weak light emanating from the cylinder gas, he fastened the film to the ceiling of the darkened test cell and exposed it stroboscopically through a transparent cylinder head for up to 23 h of running.

Daniel's findings encouraged engine designers to minimize chamber surface-to-volume ratio so that fewer HC molecules would be included in the quenched

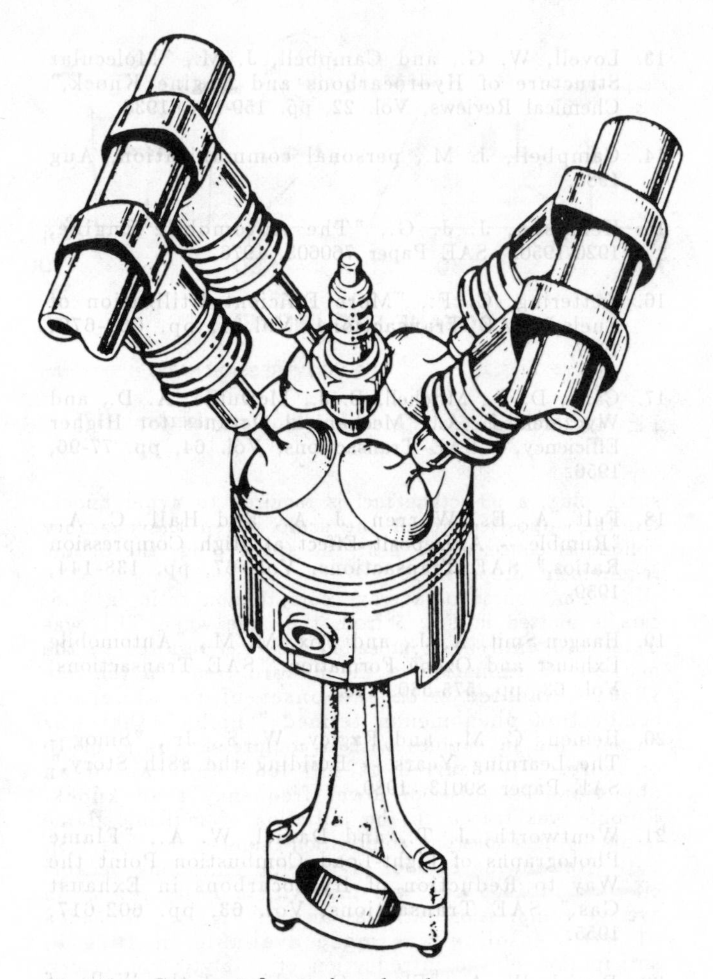

Fig. 7 Compact four-valve chamber

mixture. Later experiments by a number of investigators have raised uncertainties about the significance of the quench layer on the grounds that its contents diffuse into the bulk gas and oxidize before the exhaust process. Other causes of unburned HC associated with wall phenomena have been identified, however, which would still encourage low surface-to-volume ratios.

A schematic representation of the popular understanding of HC emissions is presented in Fig. 8. Fuel molecules absorbed by either (a) deposits on the cylinder head or (b) the oil film on the cylinder wall are desorbed as cylinder pressure falls on the expansion stroke and are not completely burned. (c) The flame may extinguish before it has consumed the fuel in the portion of the chamber most remote from the spark plug. (d) Unburned mixture is compressed into chamber crevices, the largest normally being that between the piston and the cylinder bore above the top piston ring. This transfer of unburned mixture continues until peak cylinder pressure is reached during combustion, after which the unburned mixture, shielded from combustion by the crevice, now enters the chamber too late to be completely consumed.

As suggested in Fig. 8, any HC escaping the cylinder during exhaust can still oxidize in the exhaust port and/or manifold if the exhaust temperature is high enough for the available residence time and there is sufficient oxygen present. In all U.S. cars with spark-ignition engines today, the exhaust passes on to a

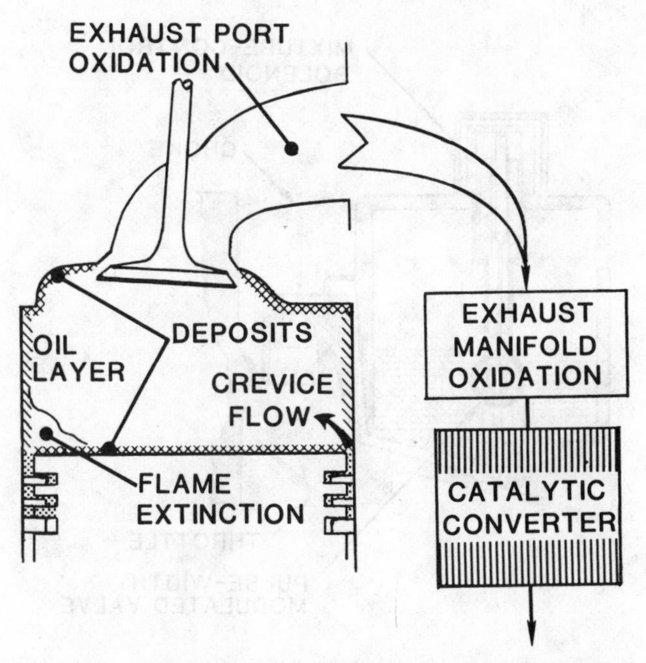

Fig. 8 Origins of unburned hydrocarbon emissions

catalytic converter, which promotes chemical reactions that transform engine-out emissions into benign gaseous products.

Production of carbon monoxide (CO), which like HC is a regulated product of incomplete combustion, is normally associated with the burning of a rich mixture. Rich mixtures are most likely to be encountered during cold starts, idle, closed-throttle decelerations, strong accelerations, and operation at heavy load.

Oxides of nitrogen (NOx) consist mainly of nitric oxide, which is an equilibrium product of combustion. Its concentration increases exponentially with temperature, and at a given temperature it is most abundant at a mixture ratio slightly leaner than stoichiometric.

Too many investigators to enumerate have contributed to this understanding of the sources of exhaust emissions, but that understanding has been crucial to the outstanding reduction in passenger-car tailpipe emissions that has been effected.

CONTROL SYSTEMS

To appreciate some of the measures taken to control exhaust emissions, some historical understanding of the ignition-timing and mixture-preparation systems is helpful.

Spark timing

For each engine speed and throttle setting, there is a spark advance that gives maximum torque. Advancing the spark from that setting, in addition to sacrificing fuel economy, produces higher maximum cylinder pressure and combustion temperature, and is thus more likely to cause combustion knock. Retarding the spark from that setting, in addition to sacrificing fuel economy, produces lower maximimum cylinder pressure and combustion temperature, as well as higher exhaust-gas temperature.

For best fuel economy, the spark needs to be advanced as engine speed increases. At a given speed the spark also needs to be advanced from its best full-throttle setting as the throttle is closed because of slower burning at part throttle. In addition, optimum timing at a given speed and throttle position depends on mixture strength. Into the 1920s, the driver was confronted with a control lever radiating from one side of the steering column with which he was expected to adjust spark timing. It seems likely that most of the time, the spark was not adjusted to its optimum setting for the engine operating condition.

Automatic spark advance was later built into the distributor. Typically, centrifugal flyweights advanced the timing with increasing engine speed, and a diaphragm exposed to intake-manifold vacuum advanced the spark as the throttle was closed.

Today a table of spark advance as a function of engine speed and manifold pressure can be stored in the on-board computer. Additional input parameters can be used to modify the table as appropriate. For example, a coolant-temperature signal can be used to adjust spark timing differently during engine warmup, or a time delay can be incorporated to accommodate the differing response rates of elements of the engine during transients. Use of a knock sensor that overrides stored timing schedules to avoid combustion knock was previously cited. Such capabilities of the computer have helped engines to comply with stringent emission standards while minimizing fuel-economy losses.

Mixture preparation

For many years the carburetor followed the basic principle illustrated in the engine of Fig. 1, namely relying on a partial vacuum created in the incoming airstream to draw fuel from a tube filled to a level controlled by a float bowl. Many embellishments were added, to be sure. For example:

(a) The simple carburetor has a natural tendency to enrich the mixture as the airflow rate through it increases because air is compressible and the fuel is not. A number of schemes were devised to overcome that characteristic.

(b) At idle, the depression in the throat of the carburetor venturi is insufficient to overcome the viscosity of the fuel it is supposed to draw into the airstream. Special idle circuits of various types were added to overcome that problem, at the same time being designed to enrich the mixture at idle in compensation for the high residual-gas dilution in the cylinder that occurs at high intake-manifold vacuums.

(c) In order to realize the total power potential of the engine at full throttle, it is necessary to enrich the mixture from the lean ratio that is preferred at cruise for good fuel economy. Using a richer mixture at full load also avoids the valve burning that might otherwise be experienced with lean mixtures. Provisions were therefore incorporated for mixture enrichment at high loads.

(d) Typically the fuel in the intake manifold is partially vaporized and carried along by the airstream, and is partially in the form of a liquid layer migrating along the manifold walls toward the cylinder. When the throttle is suddenly opened for more power, the airstream and the vaporized fraction of the fuel respond promptly, but the liquid layer cannot. Moreover, the increased absolute pressure in the manifold when the throttle is opened encourages more of the fuel to stay in the liquid layer. To avoid a sudden leaning of the mixture entering the cylinder on sudden accelerations, then, extra compensatory fuel is added to enrich the mixture temporarily.

With the introduction of the reducing catalyst for NOx control in 1981, it became necessary to control air-fuel ratio within narrower limits about stoichiometric than could be expected from such a carburetor. The result was closed-loop control of the air-fuel ratio, using an oxygen sensor in the exhaust to tell the on-board computer when the mixture ratio was drifting too far from stoichiometric. That spelled the end of the traditional carburetor, giving birth to three options in use today.

First is the electronic carburetor, shown in Fig. 9. It bears a strong resemblance to the traditional carburetor. However, fuel metering is done through a computer-controlled variable-area orifice, rather than a fixed-area orifice or one having its area varied by a metering rod mechanically linked to the throttle. The flow area in the electronic carburetor is controlled by a pulse-width-modulated solenoid.

The second option is single-point injection, typified by the throttle-body injector of Fig. 10. This involves an electronically controlled injector located at the entrance to the intake manifold, where the carburetor would normally be positioned. Fuel is delivered under pressure from a pump in the fuel tank, and is metered and atomized by the injector. The portion of the fuel delivered to the injector that is not sprayed into the intake manifold is recirculated back to the tank.

A third possibility is multi-point injection into the intake manifold. A popular approach is to inject fuel into each intake port, as illustrated in Fig. 11. Because this technique avoids wetting the intake-manifold walls over their entire length, the need for heating the intake manifold shared by other systems is obviated. This elimination of intake-manifold heating offers some gain in volumetric efficiency.

During cold starts a butterfly choke valve was once located ahead of the carburetor to enrich the air-fuel ratio in compensation for inadequate vaporization of the carbureted liquid fuel in the cold engine. In the 1920s, manual choking was the job of the driver, and opening the choke again once the engine was warm was an easily forgotten task, leading to an overly rich mixture for a warmed-up engine, high HC and CO emissions, and poor fuel economy.

After World War II, the automatic choke gradually replaced the manual version. A bi-metallic coil heated by the coolant, by the exhaust gas, or electrically was used to open the choke as the engine came up to operating temperature.

Today the mixture enrichment formerly accomplished with the choke valve is assigned to the computer, which signals the injector to increase the fuel supply by an amount dependent upon measured coolant temperature.

In fact, the computer assumes control of all the jobs once assigned to the various elements of the not-so-simple "simple" carburetor listed above. In addition, it is able to provide an air-fuel ratio that is independent of altitude, something the traditional automotive carburetor did not do.

Either of two techniques is used by the computer to determine airflow so that the appropriate fuel rate can be calculated. One involves direct measurement of airflow by an air meter located upstream of the intake manifold. The other involves a calculated airflow based on measured air temperature and intake-manifold absolute pressure, engine speed, and volumetric efficiency from a look-up table stored in the computer. The oxygen sensor in the exhaust manifold, which responds too slowly to provide the primary mixture-

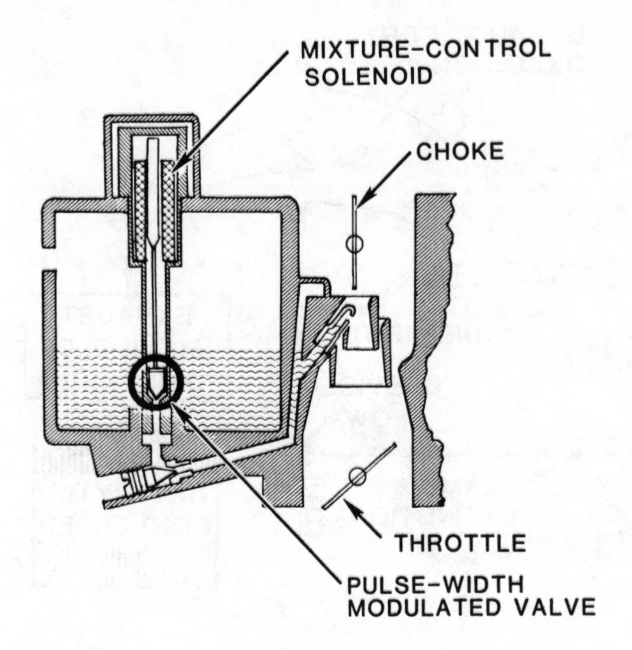

Fig. 9 Electronic carburetor

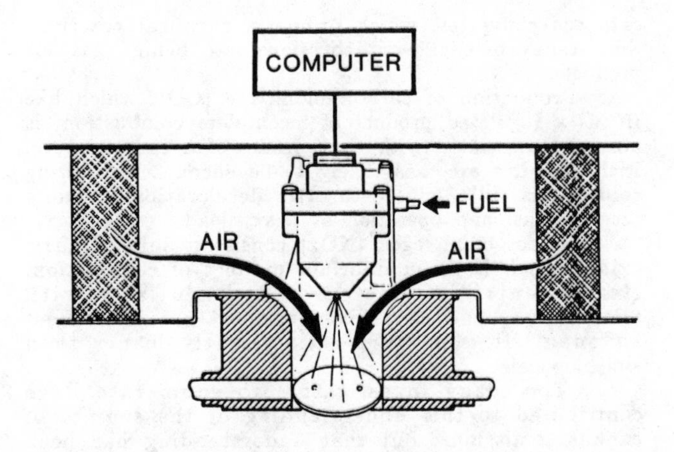

Fig. 10 Throttle-body injector

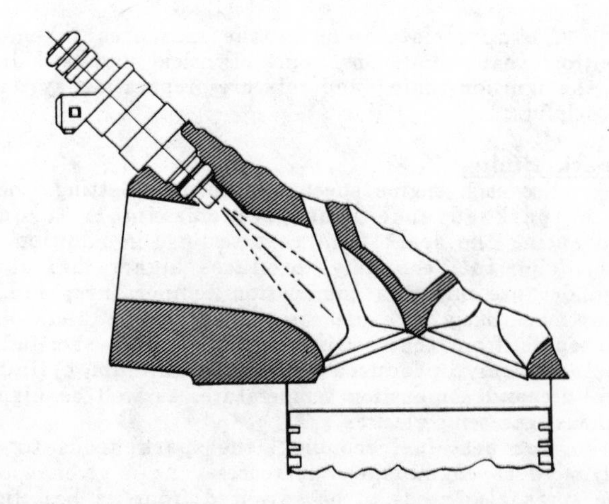

Fig. 11 Port fuel injector

control signal, calibrates whichever of these two systems is selected to compensate for control drift, changing engine characteristics, etc.

Emission controls

The first emission control to be added to the automobile was positive crankcase ventilation (PCV). Applied to all U.S. cars in 1963 (earlier in California), it eliminated the draft tube previously used to vent crankcase blowby gases to the atmosphere and ducted them to the engine intake system instead.

Nationwide regulation of tailpipe HC and CO started in 1968. Examples of measures taken to control these gases were:

(a) lowered compression ratio, which generally decreases crevice HC, and raises post-combustion gas temperature to promote oxidation of HC and CO

(b) retarded spark to raise post-combustion gas temperature

(c) increased idle speed, accompanied by a leaner idle mixture, for lower HC and CO

(d) slowed throttle closing on decelerations

(e) preheated intake air to a thermostatically controlled temperature for maintenance of a more uniform air-fuel ratio schedule under various ambient conditions.

(f) added air to the exhaust from a belt-driven pump to promote oxidation of HC and CO

It is noteworthy that most of these emissions-reduction measures penalized fuel economy.

In 1971, control of evaporative emissions was instituted nationwide. The carburetor float bowl and the air space above the fuel tank were vented to a canister containing activated charcoal. The fuel vapors trapped on the charcoal were subsequently desorbed and routed to the engine.

In 1973, NOx became regulated nationwide. Recirculating exhaust gas (EGR) into the intake manifold was found to lower NOx substantially by decreasing the combustion temperature. Because flame speed is decreased, additional spark advance is required. Some early EGR control systems contributed to deterioration of driveability, but with current systems and fast-burn combustion chambers, this has been overcome, and some fuel-economy improvement may be realized from using EGR properly.

Concurrent with nationwide regulation of NOx in 1973, HC and CO standards were incremented downward. Especially on four-cylinder engines, the air pump was replaced by Pulsair, a sel-acting pump driven by oscillations in the exhaust pressure.

In 1975, federal HC and CO standards were lowered to the point where the oxidizing catalytic converter began to take over. In the converter, small quantities of noble metal, usually platinum and palladium, are finely dispersed over a large surface area to which the exhaust gas is exposed. Above a light-off temperature, the catalyst sites promote oxidation of HC and CO to carbon dioxide and water vapor, provided a source of oxygen is present. Sometimes an engine-driven air pump or Pulsair is adopted to ensure such an oxidizing atmosphere.

In 1981, a two-thirds reduction in the federal NOx standard ushered in the reducing catalyst, with rhodium being the preferred constituent. At rich mixtures, the reducing catalyst can be quite effective, but it does not perform well in a highly oxidizing atmosphere. Consequently, dual-bed converters appeared in which the upstream converter received stoichiometric exhaust from the engine and reduced NOx, along with oxidizing some HC and CO. Then the exhaust gas passed through the downstream oxidizing converter for further removal of HC and CO. Air was injected between the two converters to ensure an oxidizing atmosphere in the second unit.

As an alternative, platinum, palladium and rhodium are all mixed in a single converter that is operated at the stoichiometric air-fuel ratio, where both the reducing and oxidizing catalysts share reasonably high conversion efficiencies over a narrow range of stoichiometry.

Because both of these approaches to NOx control demand tight control of air-fuel ratio, closed-loop control of mixture ratio, monitored by the exhaust oxygen sensor, is employed as discussed previously. Responding to the ratio of oxygen concentrations in the exhaust and the ambient environment, the oxygen sensor behaves almost like a switch that is turned on and off as the exhaust air-fuel ratio crosses the stoichiometric ratio.

SATISFYING THE CUSTOMER

The commercial success of the automobile depends on customer satisfaction. The foregoing review is fraught with examples of how the engine has been improved to make its operation more convenient, to provide better fuel economy, and to improve it environmentally. Because these advances have accrued incrementally, it is easy to lose sight of the enormous progress that has been made over the years. Consequently, a quantitative overview of gains made in specific power, fuel economy and tailpipe emissions is appropriate.

Specific power

A representative 1988 U.S. passenger-car engine delivers 112 kW (150 hp) from a displacement of 3.8 L (231 in.^{3}). At the turn of the century it would have required a displacement of 45 L (2730 in.^{3}) to develop this power!

At the bottom of Fig. 12, power is plotted versus displacement for sixteen 1900-vintage automotive engines [8]. Both axes are logarithmic, so lines of constant specific power in kW/L displacement are straight and parallel, as indicated. At the top of the plot, ranges are shown for engines powering U.S. 1988 passenger cars, both foreign and domestic, grouped according to whether they use two-valve heads, 4-valve heads, or either of these choices with turbocharging. Naturally aspirated two-valve heads are seen to have developed 20-40 kW/L. Turbocharging boosted output to 50-70 kW/L, with naturally aspirated four-valve heads falling in between.

There is a tendency for engines of a given generation and application to be limited to approximately the same mean piston speed at their rating points, one reason being that piston and ring friction dominates the friction losses and is sensitive to mean piston speed. It can be shown that the area-specific power of a four-stroke-cycle engine, i.e., the ratio of its rated power to the total cross-sectional area of its cylinders, can be expressed as

$$\frac{P}{A} = \frac{\bar{p}\,U}{4} \qquad (1)$$

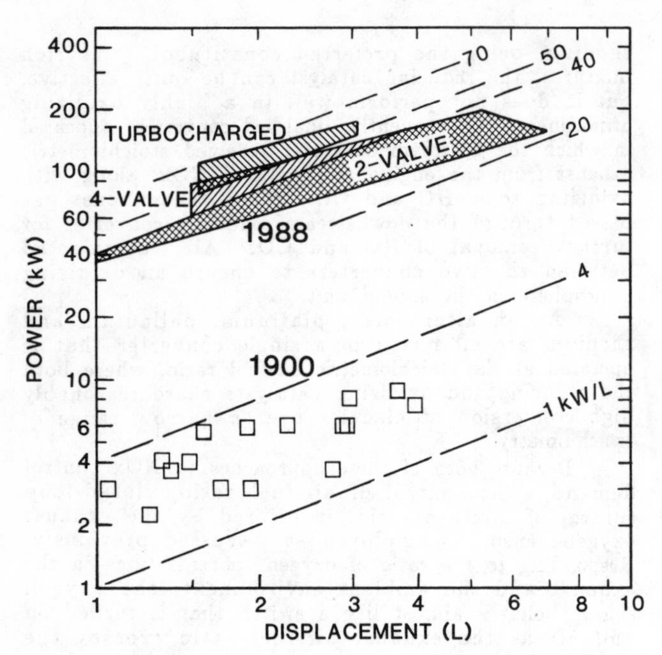

Fig. 12 Power vs. displacement, 1900 and 1988

where P = engine rated brake power

A = total cross-sectional area of cylinders

$\bar{p}$ = brake mean effective pressure

U = mean piston speed

In the lower left corner of Fig. 13, brake mean effective pressure (BMEP) is plotted versus mean piston speed (MPS) for a collection of 1900-vintage automotive engines at their rated-power conditions [8]. It is evident from Eq. (1) that on these coordinates, curves of constant area-specific power are hyperbolas. This parameter is plotted on a relative basis in Fig. 13.

In the upper right corner of the graph, circular data points are shown for a collection of 1988 passenger-car engines of one of the full-line domestic manufacturers. It is seen that the area-specific power is typically eight times that of a representative 1900 automotive engine. It is also seen that this has been

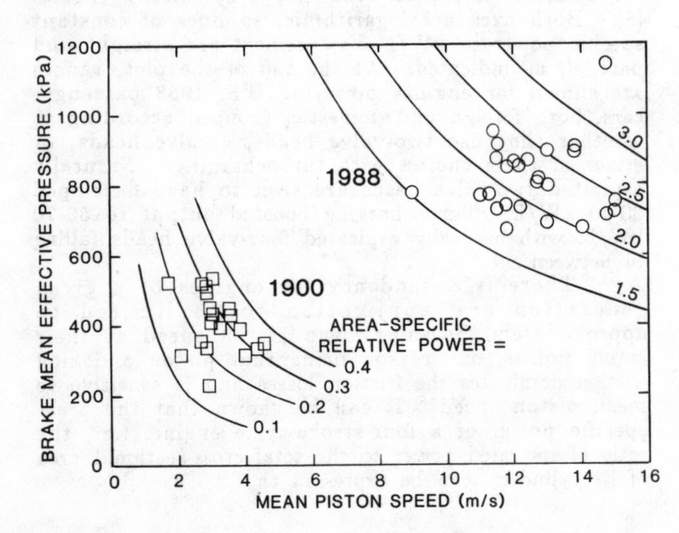

Fig. 13 BMEP vs. mean piston speed, 1900 and 1988

the result of a doubling of BMEP and a four-fold increase in MPS. Increased compression ratio, made possible by improvements in chemical and mechanical octane numbers, is responsible for a good share of the increase in BMEP. The increase in MPS is largely attributable to higher rotational speeds arising from improvements in mechanical design, lighter reciprocating parts, advances in tribology, and the availability of better lubricants.

Fuel economy

The ultimate measure of fuel economy for the automotive engine is buried in the fuel economy of the vehicle, normally measured in the U.S. in mi/gal. However, on a specified fuel the vehicle fuel economy depends not only on the engine, but also on the accessory load, the drivetrain, the mass and aerodynamic drag of the vehicle, the rolling resistance of the tires, and the driving schedule. For a fixed driving schedule, of the non-engine related factors, the most influential for cars of a given generation is generally the vehicle mass. Fuel Economy Index (FEI) has been used as a means of separating out the effect of vehicle mass from the other factors. A case can be made for the position that FEI is a simple parameter that reasonably measures the efficiency of the powertrain, which for a given level of technology in the transmission, aerodynamics and tires is dominated by the engine [23].

Fuel economy index can be defined as

$$FEI = (MPG \times TWC)/1000 \qquad (2)$$

where FEI = fuel economy index in (lb-mi/gal)/1000

MPG = mi/gal on EPA combined urban/highway schedule

TWC = EPA test weight class (lb)

This definition reflects the availability of fuel economy data from the transient EPA urban and highway tests. FEI may be thought of as the number of miles that a gallon of fuel will transport 1000 lb of vehicle mass on the prescribed driving schedule.

In earlier days, the standardized EPA tests did not exist, and vehicle fuel economy was customarily evaluated at a constant road speed. Average FEIs for sizable fleets of cars from 18 manufacturers operated at a constant speed of 64 km/h (40 mi/h) are presented in Fig. 14 for the period from 1930 to 1955 [17]. A steady upward trend, interrupted by the war years, is noted, with an improvement of about 60% being realized over this quarter-century. From the trend in average compression ratio over those years, from Fig. 5, about a 15% improvement in efficiency might be expected from that source. Additional gains are available if the extra power provided by increased compression ratio is used to improve fuel economy by reducing displacement or by increasing the final drive ratio rather than being used to increase performance.

Up to another 15% gain came from leaning out the mixture. For a fleet of ten different 1927 cars, an average air-fuel ratio of 13 was reported at a steady 64 km/h (40 mi/h), accounting for a loss of about 15% from incomplete combustion of so rich a mixture [24]. By 1940 that source of lost energy had been essentially eliminated by leaning the mixture. The balance of the gain in FEI over that time period is attributable to design improvements in the engine, drivetrain and vehicle.

In more recent times, the composite fuel economy, which assumes 55% of the distance traveled by a car is simulated by the EPA urban schedule and 45% by the EPA highway schedule, has been used to regulate

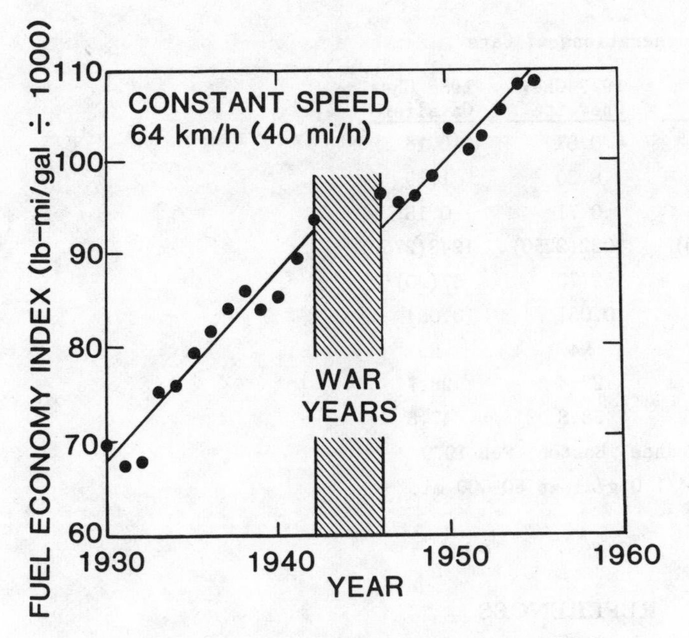

Fig. 14 FEI trend at constant speed, 1930-55 [17]

Corporate Average Fuel Economy. In Fig. 15, FEI based on this composite value is plotted against years for the first two decades of tailpipe emissions regulation. It is seen that during the early regulatory years, FEI dropped for reasons previously discussed as manufacturers struggled to comply on schedule. With widespread introduction of the oxidizing catalytic converter in 1975, some of those early engine changes that traded fuel economy for lower emissions could be reversed. FEI is seen generally to have increased at a decreasing rate since introduction of the catalytic converter.

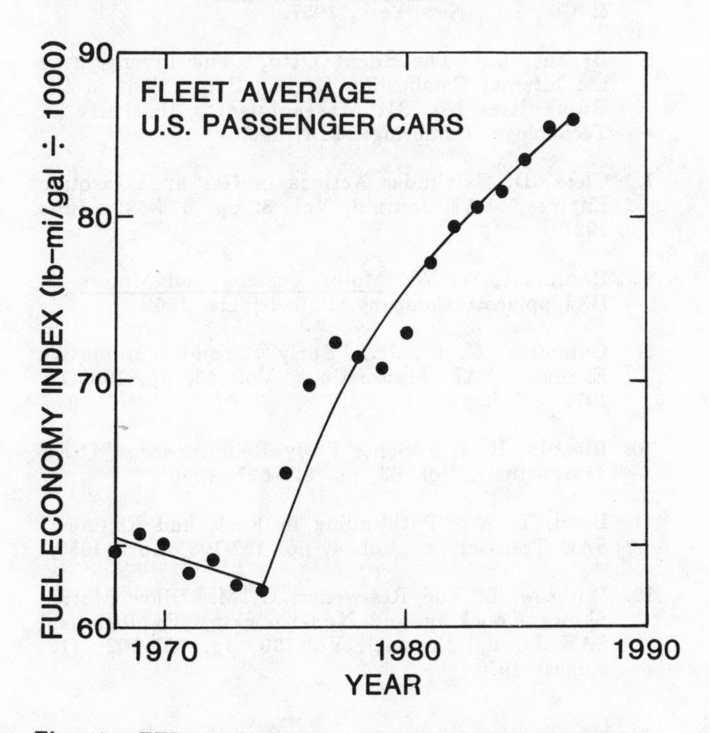

Fig. 15 FEI trend, EPA composite schedule, 1968-87

Emissions

The history of federal tailpipe emission regulation is compared to levels from an unregulated car in Fig. 16. The current standards of 0.41/3.4/1.0 g/mi HC/CO/NOx must be met after 80 000 km (50 000 mi) of driving. These levels mean that it would take 25 current-model cars to emit as much HC or CO as a single car did in pre-control days. At the same time, NOx has been reduced by 67% from the average pre-control level.

Another illustration of the remarkable progress made in emissions control is provided in Table 1, where data are compared for three different generations of cars, dating from 1921, 1979 and 1988. The 1921 model pre-dated emission control. The 1979 car employed an oxidizing catalytic converter, and catalytic control of NOx was added to the 1988 car.

Evolutionary refinements in comfort and convenience in the three cars tabulated is seen to have been accompanied by increased car mass. The ratio of rated power to car mass, a rough indicator of vehicle performance capability, increased with time. Despite this, the fuel economy also improved with time. The 1921 car had a top speed of only 68 km/h (42 mi/h) and so could not even negotiate the EPA highway schedule.

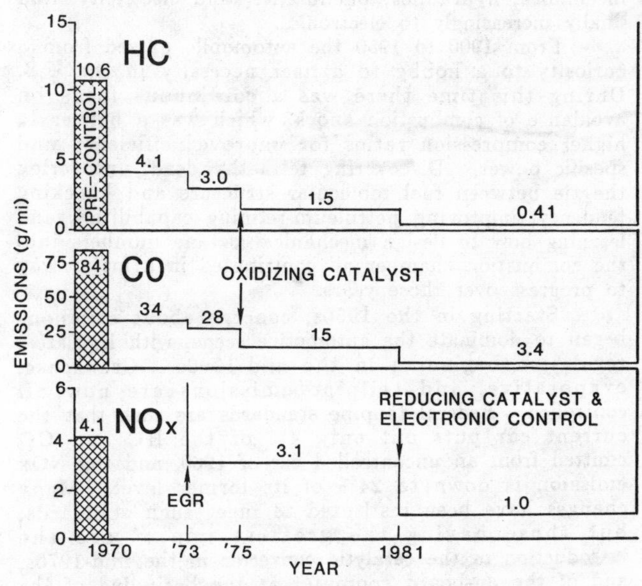

Fig. 16 U.S. tailpipe emission regulations

CONCLUDING REMARKS

It is seen from this review that the foundation for the currently dominant spark-ignition four-stroke-cycle reciprocating automotive engine had been laid by the early part of this century. What has occurred since then has been largely evolutionary. Automotive engines have been applied with 1, 2, 3, 4, 5, 6, 8, 10, 12 and 16 cylinders, in in-line, vee, and horizontally opposed arrangements. They have been both liquid- and air-cooled. A limited number of sleeve-valve engines have been used, but the poppet valve has emerged as the preferred option. Poppet valves have been actuated through pushrods, rocker arms, and directly by the cam. The camshaft has been mounted both in the block and overhead, with double-overhead camshafts also making their appearance. Regulation of

Table 1 - Comparison of Three Generations of Cars

Vehicle Characteristics	1921 Ford[a] Model T	1979 Chev[a] Chevette	1988 Chev[b] Cavalier
HC - g/mi	5.49	0.67	0.16
CO - g/mi	41.03	8.30	1.69
NOx - g/mi	2.20	0.71	0.15
Test weight class - kg (lb)	908(2000)	1022(2250)	1248(2750)
Rated power - kW (hp)	15(20)	52(70)	67(90)
Power/mass - kW/kg	0.016	0.051	0.054
Transmission	M2	M4	M5
Urban economy - mi/gal	16.7	28.4	28.7
Highway economy - mi/gal	----	38.8	47.6

[a]From Department of Transportation Conference, Boston, Feb 1979.
[b]Certification data. Emissions <0.41/3.4/1.0 g/mi at 50 000 mi.

the engine has passed increasingly from manual control by the driver, to automatic systems relying on mechanics, hydraulics, pneumatics, and electricity, and finally increasingly to electronics.

From 1900 to 1950 the automobile passed from a curiosity to a hobby to a near necessity in the U.S. During this time there was a continuous focus on avoidance of combustion knock, which was a barrier to higher compression ratios for improved efficiency and specific power. Discovering tetraethyl lead, uncovering the tie between fuel molecular structure and knocking tendency, improving petroleum-refining capabilities, and learning how to design mechanical octane numbers into the combustion chamber all contributed in a major way to progress over those years.

Starting in the 1950s, concern about emissions began to dominate the automotive scene, with legislative regulation beginning in the mid-1960s. Crankcase, evaporative, and tailpipe emissions are now all controlled. Federal tailpipe standards are such that the current car puts out only 4% of the HC and CO emitted from an uncontrolled car of 1960, and the NOx emission is down to 24% of its former level. Many changes have been instituted to meet such standards, but those having the greatest impact are the introduction of the catalytic converter in the mid-1970s, and of the on-board computer at the beginning of the 1980s.

Federal fuel-economy standards were set in 1974. Since that time the EPA composite fuel economy of the average U.S. passenger car has about doubled, although not all of this has come from the engine alone. If Fuel Economy Index is accepted as an indicator of powertrain efficiency, a 50% improvement has been realized by the average U.S. passenger car since 1974.

The improvements that have been made in engine specific power and fuel economy, accompanied by an enormous reduction in emissions, would undoubtedly amaze the early engine pioneers, and the end is not yet in sight.

REFERENCES

1. Doyle, G. R., The World's Automobiles, 1880-1958, Temple Press Limited, London, 1959.

2. Wren, J. A., Godshall, J. I., Kollins, M. J., Wagner J. K., and Yanik, A. Y., "The Automobile - the Unwanted Child," SAE Paper 890812, 1989.

3. Nevins, A., Ford: The Times, the Man, the Company, Charles Scribner's Sons, New York, 1954.

4. Kettering, C. F., and Orth, A., The New Necessity, Williams and Wilkins Company, Baltimore, 1932.

5. Boyd, T. A., Professional Amateur, E. P. Dutton & Co., Inc., New York, 1957.

6. Bryant, L., "The Silent Otto," The Invention of the Internal Combustion Engine, Publications in the Humanities No. 81, Massachusetts Institute of Technology, Cambridge, MA, 1966.

7. Clerk, D., "Cylinder Actions in Gas and Gasoline Engines," SAE Journal, Vol. 8, pp. 523-539, June 1921.

8. Beaumont, W. W., Motor Vehicles and Motors, J. B. Lippincott Company, Philadelphia, 1900.

9. Cummins, C. L., Jr., "Early IC and Automotive Engines," SAE Transactions, Vol. 85, pp. 1960-71, 1976.

10. Ricardo, H. R., "Some Early Reminiscences," SAE Transactions, Vol. 63, pp. 620-627, 1955.

11. Boyd, T. A., "Pathfinding in Fuels and Engines," SAE Transactions, Vol. 4, pp. 182-195, April 1950.

12. Withrow, L., and Rassweiler, G. M., "Slow Motion Shows Knocking and Non-Knocking Explosions," SAE Journal (Trans.), Vol. 39, pp. 297-303, 312, August 1936.

13. Lovell, W. G., and Campbell, J. M., "Molecular Structure of Hydrocarbons and Engine Knock," Chemical Reviews, Vol. 22, pp. 159-174, 1938.

14. Campbell, J. M., personal communication, Aug 1980.

15. Hempson, J. J. G., "The Automobile Engine, 1920-1950," SAE Paper 760605, 1976.

16. Kettering, C. F., "More Efficient Utilization of Fuels," SAE Transactions, Vol. 1, pp. 669-679, 1947.

17. Caris, D. F., Mitchell, B. J., McDuffie, A. D., and Wyczalek, F. A., "Mechanical Octanes for Higher Efficiency," SAE Transactions, Vol. 64, pp. 77-96, 1956.

18. Felt, A. E., Warren, J. A., and Hall, C. A., "Rumble -- A Deposit Effect at High Compression Ratios," SAE Transactions, Vol. 67, pp. 138-144, 1959.

19. Haagen-Smit, A. J., and Fox, M. M., "Automobile Exhaust and Ozone Formation," SAE Transactions, Vol. 63, pp. 575-580, 1955.

20. Heinen, C. M., and Fagley, W. S., Jr., "Smog - The Learning Years -- Building the 88th Story," SAE Paper 89013, 1989.

21. Wentworth, J. T., and Daniel, W. A., "Flame Photographs of Light-Load Combustion Point the Way to Reduction of Hydrocarbons in Exhaust Gas," SAE Transactions, Vol. 63, pp. 602-617, 1955.

22. Daniel, W. A., "Flame Quenching at the Walls of an Internal Combustion Engine," Sixth Symposium (International) on Combustion, pp. 886-892, 1957.

23. Amann, C. A., "The Powertrain, Fuel Economy and the Environment," International Journal of Vehicle Design," Vol. 7, pp. 1-34, Jan/March 1986.

24. Lovell, W. G., Campbell, J. M., D'Alleva, B. A., and Winter, P. K., "A 13-Year Improvement in Mixture Ratios," SAE Journal (Trans.), Vol. 48, pp. 160-164, 1941.

AN HISTORICAL OVERVIEW OF EMISSION-CONTROL TECHNIQUES FOR SPARK-IGNITION ENGINES: PART A — PRIOR TO CATALYTIC CONVERTERS

J. R. Mondt
AC Rochester Division
General Motors Corporation
Flint, Michigan

ABSTRACT

Starting in the 1960's, spark-ignition engines have changed markedly in response to requirements for lowered exhaust and evaporative emissions. Specific emission requirements are set forth in Federal Regulations. Congress passed the Clean Air Act in 1963, and in 1965 added an amendment to regulate emissions from automobiles. Thus, automotive engineers embarked into an era of increased emphasis on engineering for control of emissions. In addition to changes made to basic engine designs, such as lowered compression ratios, reduced valve overlaps, and modified combustion chambers, substantial changes have been made in engine controls. Engine control systems included spark timing control by ported and manifold vacuum using, for example, TVS (thermal vacuum switches), TVV (thermal vacuum valves), VDV (vacuum delay valves), and VMV (vacuum modulator valves). The emission-control systems also include PCV (positive crankcase ventilation), EGR (exhaust gas recirculation), ECS (evaporation-control system), THERMAC (thermostatic air cleaner), AIR (air injection reactor), PAIR (Pulsair air injection reactor), and EFE (early fuel evaporation). These systems were developed to meet emission requirements prior to the introduction of catalytic converters in 1975.

INTRODUCTION

Congress passed the Clean Air Act of 1963, and two years later added provisions for control of automobile emissions. Adding legislated demands for emission control to the marketplace demands for fuel economy and driveability challenged the automotive industry as never before. Response to this challenge has been control of the engine by devices which limit emissions from the engine. Improving old devices and adding new devices required new technology and new terminology, all synthesized as the "emission-control system."

First, this paper delineates sources of emissions from spark ignition engines in passenger cars and summarizes the history of mandated emission standards. Then, it describes the emission-control systems and devices that were developed to meet increasingly stringent emission levels through 1974.

SOURCES OF REGULATED EMISSIONS

Currently, emissions of three gaseous constituents in the exhaust from a vehicle are regulated. These constituents are unburned hydrocarbons (HC), carbon monoxide (CO), and oxides of nitrogen (NOx). Prior to emission controls, the overall relative proportions of these three constituents for a vehicle were as shown in Fig. 1 [General Motors Corp., 1971]. For a typical vehicle, 20% of the unburned hydrocarbons was traced to vapors vented from the crankcase and 20% was traced to evaporation from two sources, the carburetor and the fuel tank. The engine exhausted the remaining 60% of the unburned hydrocarbons, as well as 100% of both carbon monoxide and oxides of nitrogen.

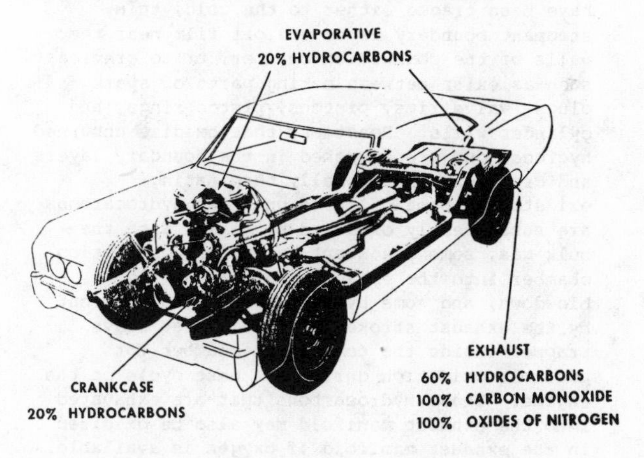

Fig. 1 - Sources of Automotive Emissions

Furthermore, these emissions in the exhaust were all produced by many chemical reactions resulting from combustion of hydrocarbons in the engine. For a typical spark-ignition engine, the combustion process produces emissions which vary with air-fuel ratio (A/F) as shown in Fig. 2 [General Motors Corp., 1971].

A typical brake specific fuel consumption (bsfc) curve is also shown in Fig. 2. The A/F for minimum bsfc varies with the specific design of an engine [Obert, 1950], but usually occurs between 16 and 18. Fortunately, emissions of both HC and CO are usually very low at the A/F for minimum bsfc. Conversely, NOx production is near its maximum at the A/F for minimum bsfc [Amann, 1980].

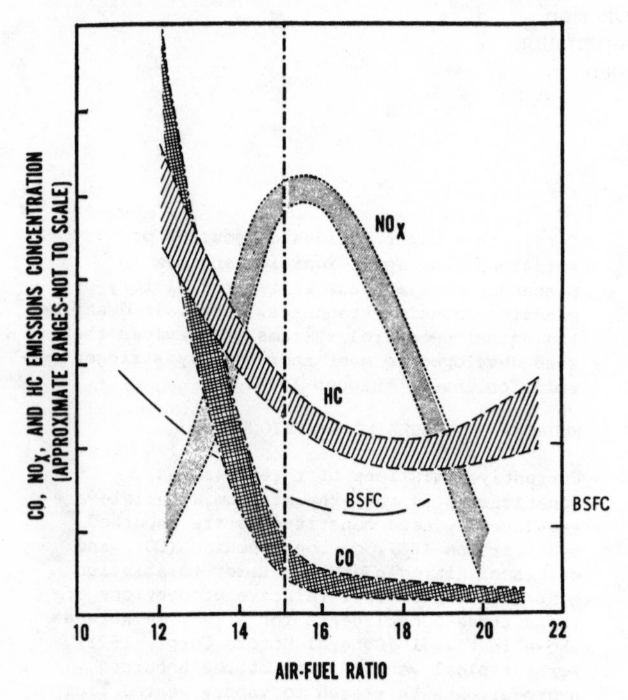

Fig. 2 - Effect of Air-Fuel Ratio on Emissions

Unburned Hydrocarbons (HC)

Most of the unburned hydrocarbons emitted from a fully warm engine running at lean A/F have been traced either to the cold, thin, stagnant boundary layer and oil film near the walls of the combustion chamber, or to crevices such as exist between mating parts of spark plugs, valve rims, pistons, piston rings, and cylinder walls. Reactions that oxidize unburned hydrocarbons are quenched in the boundary layers and crevices, thus locally terminating oxidation. Some of these unburned hydrocarbons are subsequently oxidized by mixing with the bulk gas, some are swept out of the combustion chamber into the exhaust manifold during blowdown, and some hydrocarbons are pumped out by the exhaust stroke. The remainder stays trapped inside the combustion chamber for possible oxidation during the next cycle of the engine. Those hydrocarbons that are exhausted into the exhaust manifold may also be oxidized in the exhaust manifold if oxygen is available, if temperature exceeds 750 C, and if the gases are retained in the hot environment for a sufficient residence time [Herrin, 1978]. These oxidizing reactions are called "gas-phase reactions."

As suggested by Fig. 2, when the mixture is leaned excessively, the HC levels increase because of incomplete combustion. Either the mixture is too lean to support combustion or the temperature is too low, or both. The "lean limit" is associated with degeneration of combustion and a resulting sharp increase in HC from the engine [Amann, 1980].

During a cold start, unburned hydrocarbons result from excess fuel or fuel enrichment in the combustion chamber.. This fuel enrichment is needed when the engine is cold so enough volatile fuel is present in the cylinder to initiate combustion. An overall rich A/F produces high levels of HC, as shown in Fig. 2. For a typical 1982 vehicle started at an ambient temperature of 20 C, 50% of the unburned HC leaving the tailpipe during the entire 7.5 mi. driving schedule would be discharged during the first cycle of the 1978 Federal Test Procedure (78 FTP). The 1978 FTP is described in a subsequent section and Fig. 3 of this paper, and emissions for all systems have been "normalized" to the 1978 FTP.

Carbon Monoxide (CO)

As shown in Fig. 2, high concentrations of CO are discharged from an engine operating at rich A/F. As A/F approaches the stoichiometric value, CO approaches zero, but never quite drops to zero. Some CO persists even at lean overall A/F because of cylinder-to-cylinder distribution differences, and also because there is not sufficient time to reach equilibrium of oxidation of CO to CO_2.

To illustrate the role of maldistribution, consider a six-cylinder engine with one cylinder operating slightly rich and five cylinders operating slightly lean. The five lean cylinders can make the overall A/F lean even though some CO will be present from the one rich cylinder.

As for CO resulting from non-equilibrium chemistry, CO is an intermediate product in the oxidation reaction of hydrocarbons [Herrin, 1978]. As discussed previously, unburned hydrocarbons are present even in lean mixtures, and the partial oxidation of some of the unburned hydrocarbons can produce CO.

Much CO is also produced during a cold start because of fuel enrichment. During combustion of a rich mixture, the absence of sufficient oxygen arrests the oxidation of carbon at CO instead of CO_2. Also, the partial oxidation of HC contributes to the cold-start CO. For a typical 1982 vehicle, 60% of the CO leaving the tailpipe on the 78 FTP would be discharged during the first cycle of the 78 FTP.

Similar to HC, CO has an opportunity to oxidize further to CO_2 in the exhaust manifold if there is sufficient oxygen, high enough temperature, and a long enough residence time. CO requires a slightly higher temperature for gas-phase reactions than HC, making CO slightly more difficult to oxidize [Herrin, 1978].

Large quantities of CO are also produced when mixture enrichment to low A/F is required for high power output from the engine. Some fuel enrichment to produce high power is required for fast accelerations or for operation in mountains or hilly terrain, especially if a vehicle is heavily loaded or towing a trailer. High power output is usually not required to operate a vehicle according to the 78 FTP.

Oxides of Nitrogen (NOx)

Nitric oxide (NO) is the predominate oxide of nitrogen formed during the combustion process, although some nitrogen dioxide (NO_2) may also be present. For mass emissions from automotive powerplants, the NO is assumed to be oxidized to NO_2, is added to the NO_2 present, and the sum is reported as NOx.

Nitrogen oxidized by oxygen in the air produces NO, mostly at the peak temperature during the combustion event. Since production of NO is very sensitive to peak temperature of combusion, output of NO from an engine tends to correlate with engine load, i.e., high load tends to produce large quantities of NO. Kinetics for oxidation of NO to NO_2 are slow relative to the engine cycle. Thus, most of the oxide of nitrogen exhausted from the tailpipe is NO. Eventually this NO oxidizes to NO_2 in the atmosphere.

NOx output from an engine is also very sensitive to A/F, as shown in Fig. 2, with the peak production occurring at an A/F slighly lean from stoichiometric. Unfortunately, the A/F for maximum NOx production is usually close to both the A/F for minimum HC production, and the A/F for low CO production.

Because NO production is associated with high temperatures, very little NOx is produced during the cold start, in contrast to the large HC and CO outputs from a cold start.

FEDERAL EMISSION STANDARDS

U. S. Congress passed the Clean Air Act in 1963 and amended it in 1965 to include emissions from automobiles [General Motors Corp., 1972,73,76,81]. Passenger cars began meeting regulated emission standards in 1966 in California and nationwide in 1968. Further amendments were added to the Act in 1967, 1970, and 1977. Since 1977 the Act remained unchanged until the mandatory expiration on September 30, 1981. This Clean Air Act was extended until Congress agrees on revisions. As of July 1, 1989, revisions are still being considered.

Increasingly stringent emission levels for HC, CO,and NOx are highlighted in Table I, showing Federal exhaust emission standards for the years 1970, 1973, 1975, and 1981-89, respectively. These exhaust emissions are discharged from the tailpipe of a vehicle while driving the urban schedule mandated by the 1978 FTP. Historically, there have been three driving schedules. However, all standards shown in Table I have been adjusted to reflect emissions for vehicles driven according to the current Urban Schedule, shown graphically in Fig. 3. As shown in Table I, emissions outputs in 1981 were reduced by 96, 96, and 76% for HC, CO, and NOx, respectively, relative to a typical "uncontrolled" vehicle of 1960 vintage.

TABLE I

Federal Exhaust Emission Requirements
1975 Test Procedure
(grams/mile)

	(1960)*	1970	1973	1975	1981-89	
HC	(10.6)*	4.1	3.0	1.5	0.41	(0.41)**
CO	(84)*	34.0	28.0	15.0	3.4	(7.0)**
NOx	(4.1)*	---	3.1	3.1	1.0	(0.7)**

Emission Reductions Based on Federal Standards
(percent)

HC	---	61	72	86	96
CO	---	60	67	82	96
NOx	---	--	24	24	76

*uncontrolled
**California

Also shown in Table I are California emission standards for 1981-89. California standards are lower than Federal standards for all years since 1966, except for CO in 1981-89. Because of smog, California has agressively pursued lowered emissions, implemented by its state agency, the California Air Resources Board (CARB).

As shown in Fig. 3, the EPA Urban Driving Schedule includes three main components: a cold start, a stabilized period and a hot restart. Time required to complete this driving schedule totals 41.5 minutes, including a 10-minute period with engine off just prior to the hot restart. The vehicle is actually driven 17.6 km (11 miles) on a chassis dynamometer and stops 23 times. However, the emissions from the cold start are averaged with emissions from the hot restart by combining 43% of the cold-start emissions with 57% of the hot-restart emissions. Combined emissions from the starts are added to those from the stabilized period. Thus, the effective length of the trip is only 12 km (7.45 mi).

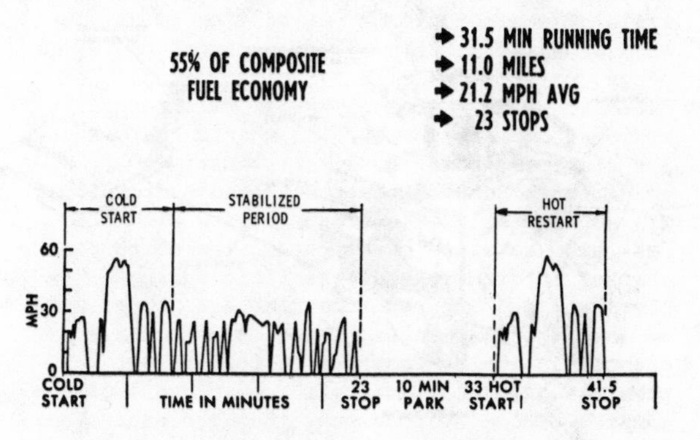

Fig. 3 - 1978 FTP Urban Driving Schedule

Vehicles are required to meet the Federal emission standards at 80,000 km (50,000 mi). A durability car must be driven 80,000 km with the emission control system installed with only maintenance scheduled by the manufacturer, or unscheduled maintenance approved by the EPA, being allowed. Emissions data points from the vehicle for the mileage between 6400 km (4000 mi) and 80,000 km (50,000 mi) are regressed linearly, and the emissions at 80,000 km are ratioed to the emissions at 6400 km to determine deterioration factors (DF) for each regulated constituent. Then, additional data cars are tested for 6400 km. The emissions measured on these cars at 6400 km multiplied by the DF cannot exceed the emission standards for any "engine/vehicle family."

EMISSION CONTROL SYSTEMS

As the standards for emissions became more stringent, emission controls became more sophisticated. Thus, chronological structure for this paper seems logical, so the use of emission-control systems has been subdivided into three historical eras:

1. Precatalytic converter, prior to 1975
2. Oxidizing catalytic converter, 1975 through 1980
3. Three-way catalytic converter, 1981 through 1989

This paper summarizes precatalytic emission control, prior to 1975.

Precatalytic Converter Era

1971 Emission Controls: A typical emission-control system for the precatalytic converter era was a 1971 system shown schematically in Fig. 4 [General Motors Corp, 1971], although AIR (Air Injection Reactor) was introduced in California in 1966. Included in the 1971 system were controls for emissions from crankcase venting, fuel evaporation, and engine exhaust. Primary control techniques were:

1. Positive crankcase ventilation (PCV);
2. Evaporation-control system (ECS);
3. Air preheat by a thermal air cleaner (THERMAC);
4. Spark control: transmission controlled spark (TCS) and a thermovacuum switch (TVS).

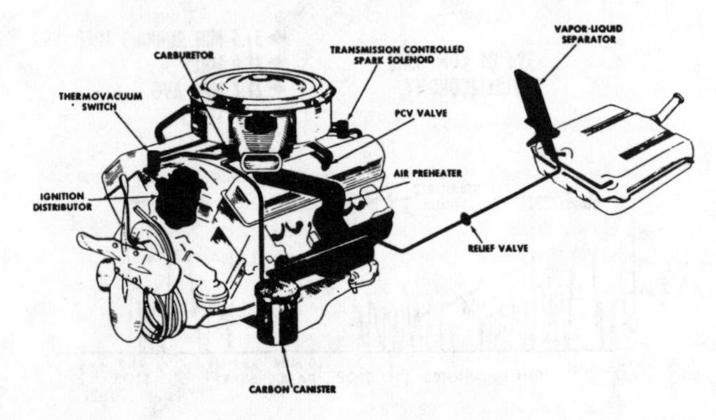

Fig. 4 - 1971 Emission-Control System

Positive Crankcase Ventilation. Positive crankcase ventilation (PCV) was employed on California cars in 1961 models and nationwide in 1963 models. PCV systems were installed prior to enactment of any state or federal regulations. Crankcase vapors contain blowby gases and evaporated oil which are mostly unburned hydrocarbons. As shown in Fig. 5, these unburned hydrocarbons are ducted to the intake manifold and added to the mixture of fuel and air entering the cylinders [General Motors Corp., 1972]. Air to replace the crankcase vapors is filtered in the air cleaner before being ducted to the crankcase, as shown in Fig. 5.

The check valve in the PCV system controls the flow of gases by allowing flow of crankcase vapors into the intake manifold, but blocking flow of combustible mixtures of fuel and air into the crankcase.

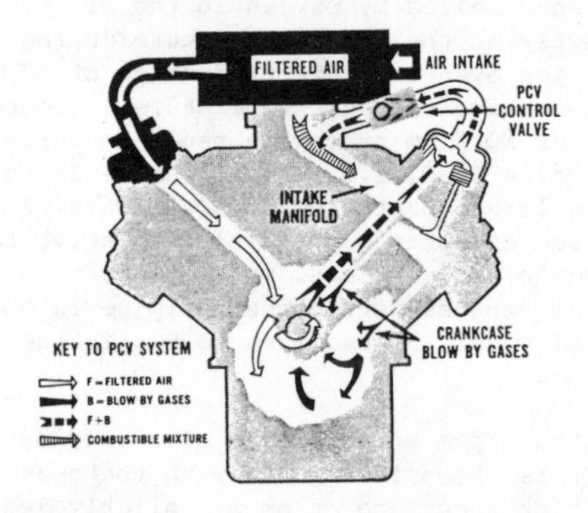

Fig. 5 - Closed PCV System

Evaporation-Control System. Systems to control discharge of evaporated fuel into the environment have been installed on passenger cars since 1970 models in California and since 1971 models nationwide. The evaporation control system (ECS) controls evaporated hydrocarbons by trapping vapors in a canister of activated charcoal. As shown in Fig. 4, the carburetor and fuel-tank vents are ducted to the charcoal canister so that vapors must pass through the charcoal to reach the atmosphere. Unburned hydrocarbons, except nonreactive hydrocarbons such as methane, are absorbed by the carbon, mostly after the engine is turned off.

To recycle the charcoal canister after the engine is started, fresh air is drawn through the charcoal to desorb previously trapped HC and to carry them into the intake manifold. With this system, evaporative losses have been lowered by 96% based on measurements from an overnight "shed" test. This shed test is specified by Federal regulations.

Preheat of Carburetor Air. Temperature of the air entering a carburetor can be controlled by an air preheat system. Air preheat can be employed both during warmup of the engine and during warmed-up operation of the engine. During a cold start, liquid fuel droplets enter the combustion chamber from the cold intake manifold. Poor combustion of these droplets cause low fuel economy and high emissions of HC and CO [Pao, 1982]. Preheating the air entering the carburetor shortens the time until the carburetor and intake manifold are warm enough to evaporate all the fuel.

During warmup, heated air is supplied to the carburetor by passing this air over a portion of the exhaust manifold. This air, warmed by the exhaust manifold, is ducted to the carburetor inlet. A thermal air cleaner (THERMAC) controls the temperature of the air entering the carburetor by mixing heated air with unheated air. The mix of hot and cold air is modulated by a vacuum-operated valve in the snorkel of the air cleaner.

As shown in Fig. 6 [General Motors Corp., 1972], a sensor inside the air cleaner senses air temperatures and controls the mix of hot and cold air to achieve and maintain a temperature of 38 C (100 F). Maintaining a constant carburetor temperature minimizes excursions in air-fuel ratio, thereby lowering emissions and increasing fuel economy.

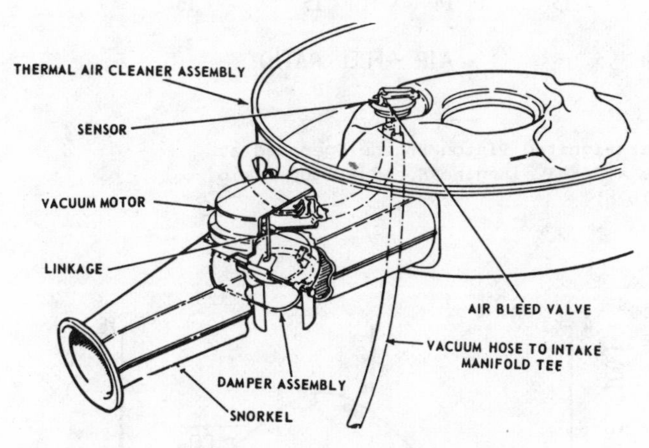

Fig. 6 - Thermac, Thermal Air Cleaner

Control of Spark Timing. Adjustments of spark timing can be used to accomplish many desirable effects, such as faster warmup, lowered emissions, improved efficiency, improved driveability, and higher exhaust temperatures. Unfortunately, one spark setting does not optimize all of the desirable effects. Furthermore, a spark setting for best efficiency varies with engine speed and load. Maximum fuel efficiency for an engine occurs when the spark is set for MBT, minimum advance for best torque. Fig. 7 illustrates a spark timing schedule as a function of speed and load for a typical 1965 vehicle. For this vehicle, the spark advance varies from a minimum of 10°BTC, before top dead center, to a maximum as high as 58°BTC. Spark advance was modulated by combining increased advance with engine speed from centrifugal flyweights with increased advance for part-load operation from manifold vacuum.

Retarding the spark setting from MBT lowers the peak cylinder pressure, lowers the peak combustion temperature, and increases the exhaust gas temperature. As a result, exhaust constituents from an engine are influenced by

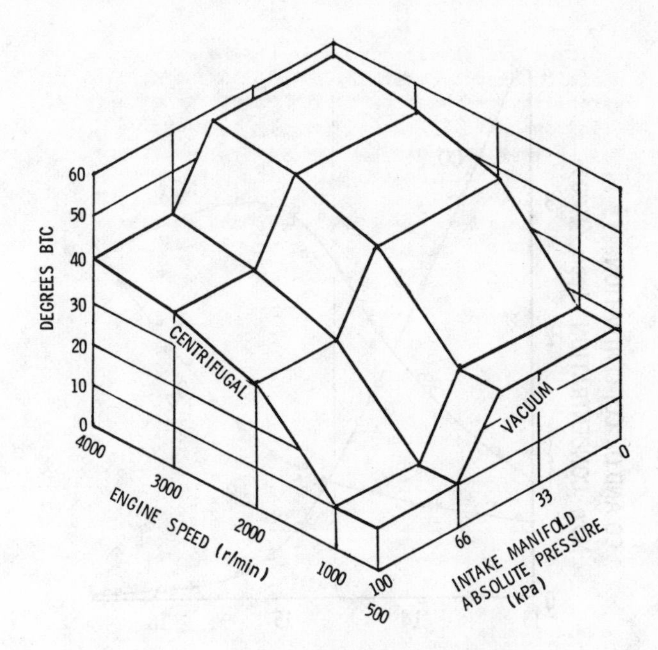

Fig. 7 - Spark Calibration for a Typical 1965 Vehicle

the spark setting. Fig. 8 and 9 depict the exhaust constituents measured for an engine-dynomometer installation of a 5.7 L spark-ignited piston engine operated at constant speed and constant exhaust flow. By controlling fuel flow, the air-fuel ratio was varied from 13 to 16. Exhaust constituents of CO, O_2, NOx, CO_2, and HC were measured at two different spark settings of 44° and 24°BTC. In these figures, HHC depicts unburned hydrocarbons measured with a heated sample line. On the other hand, CHC represents unburned hydrocarbons measured using an unheated or cold sample line. Because of less hangup of hydrocarbons in the sample line, the heated sample, HHC, indicated HC 3-5% higher than that for the unheated sample line.

Comparing NOx measured for 44°BTC spark advance, shown in Fig. 8, with NOx measured for 24°BTC spark advance, shown in Fig. 9, reveals that on this test, retarding the spark 20° lowers peak NOx about 60%. Comparing unburned hydrocarbons for the two different spark settings shows HC lowered 30% by retarding the spark 20°. Only slight lowering of CO and O_2 was observed for 20° spark retard. Unfortunately, a fuel economy penalty of approximately 10% was observed when the spark was retarded 20°.

For fully warm engine operation, these data show that retarding the spark lowers NOx drastically and lowers HC somewhat. In addition, retarded spark increases temperatures of the exhaust gases. Thus, spark retard can be combined with high flow from fast engine speed to hasten warmup of an engine and exhaust system.

As shown in Fig. 4, for 1971, transmission controlled spark (TCS) and a thermovacuum switch (TVS) were used to control spark advance. The TCS solenoid simply blocked vacuum advance from

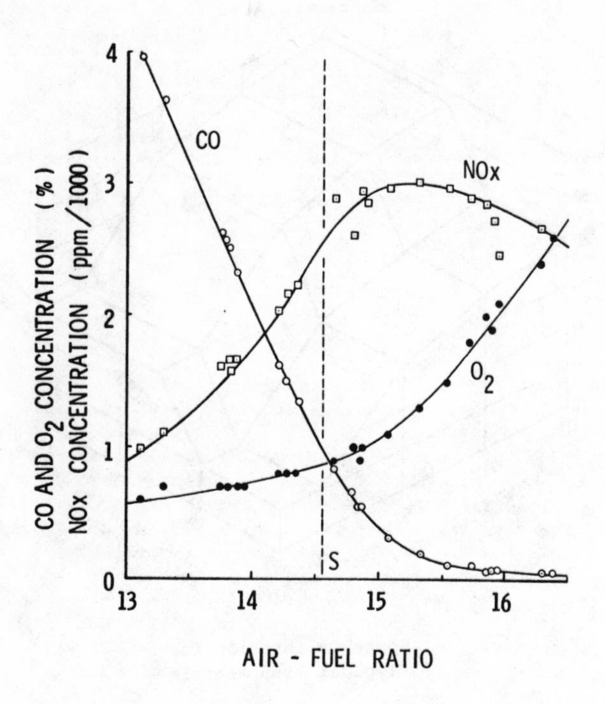

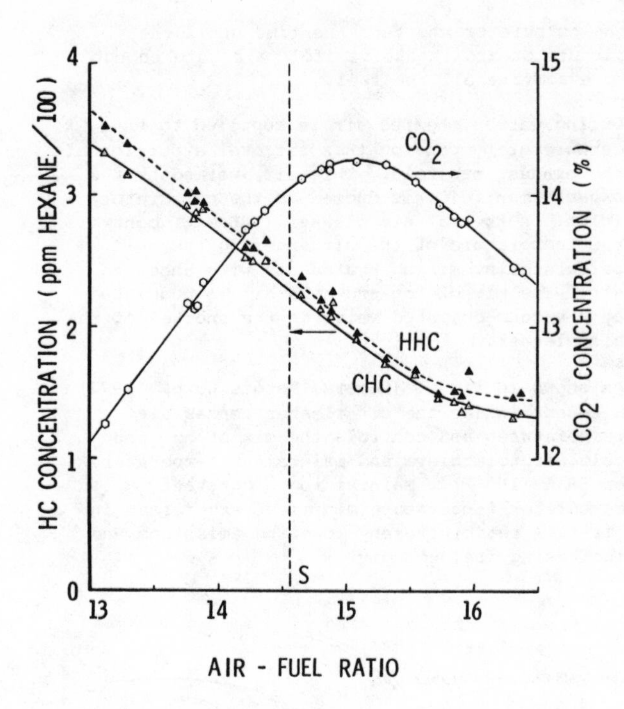

Fig. 8 - Exhaust Emissions from a Spark-Ignited Piston Engine Operated at Part Throttle, Spark Advance = 44°BTC [Engine Speed = 1600 r/min, Total Flow = 140 kg/h (310 lb/h)]

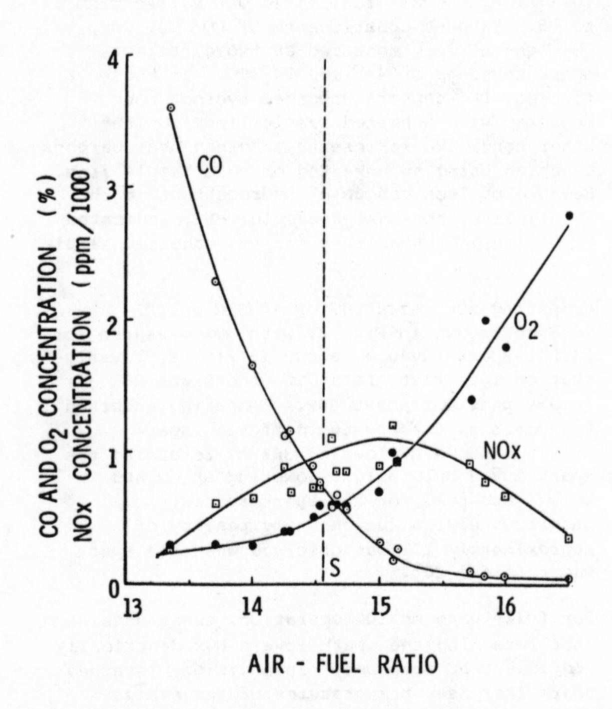

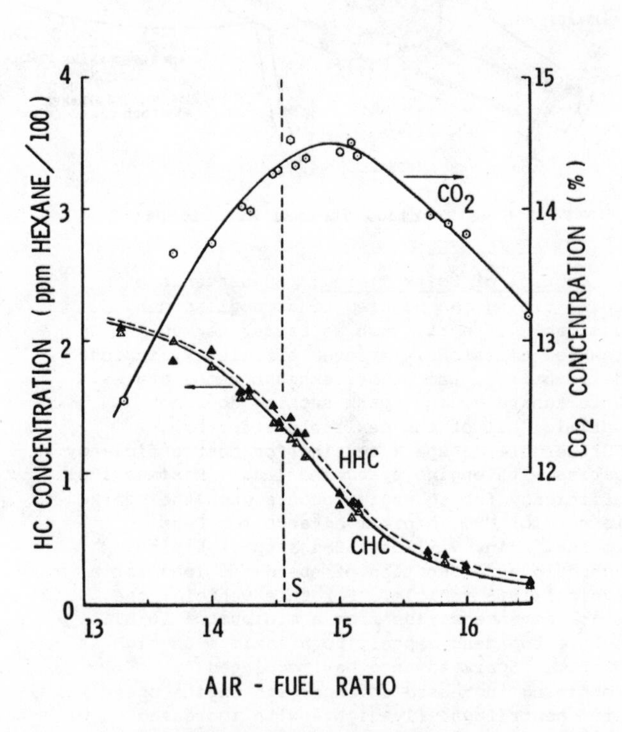

Fig. 9 - Exhaust Emissions from a Spark-Ignited Piston Engine Operated at Part Throttle, Spark Advance = 24°BTC [Engine Speed = 1600 r/min, Total Flow = 140 kg/h (310 lb/h)]

the distributor except when the transmission was shifted to high gear. Thus, spark was retarded except for vehicle operation in high gear, thereby incurring a penalty in fuel economy for all operation in lower gear ranges. If retarding the spark caused overheating of the engine, the TVS would open a vacuum line which bypassed the TCS solenoid to reinstate vacuum advance.

Closure for 1971 Emission Controls. Thus by 1971 the emission control systems included control of crankcase vapors, control of evaporation to the atmosphere, and control of HC and CO by engine tuning. Quicker warmup was achieved by an air preheat system, retarded spark timing, and fast idle speed.

1973 Emission Controls: As shown in Table I, allowable levels for HC and CO emissions lowered somewhat in 1973. Also, NOx control was required in 1973 to meet a standard of 3.1 g/mi, lowered from an uncontrolled level of 5.0 g/mi. As shown in Fig. 10 [General Motors Corp., 1973], a typical 1973 emission-control system contained an air injection pump and an exhaust gas recirculation (EGR) valve.

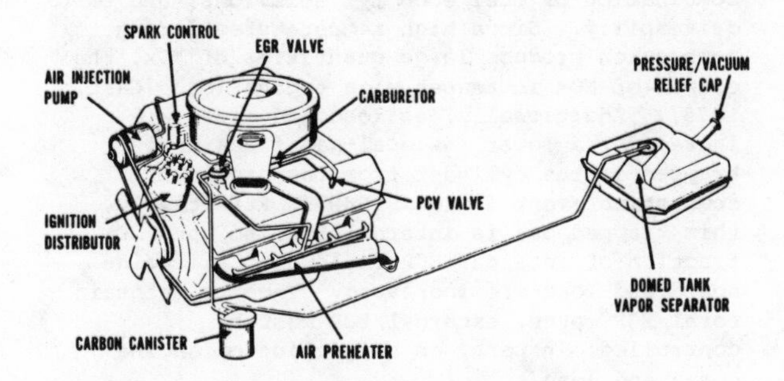

Fig. 10 - 1973 Emission-Control System

Air Injection Systems. Additional air from a rotary-vane air pump, shown in Fig. 10, was injected into the exhaust ports upstream of the exhaust manifolds to promote oxidation of CO and HC remaining in the effluent from the cylinders. Thus the added air in the exhaust manifolds promoted thermal reactions, and the system was called Air Injection Reactor (AIR) [King et al., 1970].

An alternative technique to supply additional air into the exhaust manifold, Pulsair [Gast, 1975], utilized inherent exhaust pressure pulses. The Pulsair injection reactor (PAIR) uses a quick-acting Pulsair check valve, shown in Fig. 11, to admit air into the manifold when the pressure oscillates to subatmospheric levels and to block flow out of the manifold when the pressure inside the manifold oscillates to levels above atmospheric. Pulsair is more effective with fewer cylinders, hence was included with the four-cylinder engine in the Cosworth Vega in 1975 models.

Pulsair requires no useful power, which is not the case for a belt-driven pump. Also the air supply rates for Pulsair match the requirements to oxidize HC and CO from the engine better than

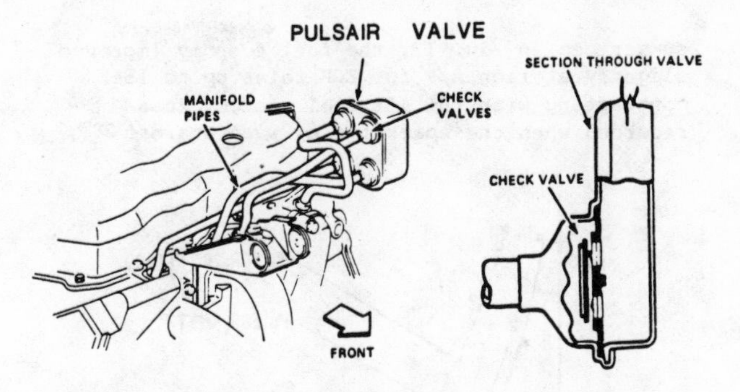

Fig. 11 - Pulsair

the air supply rate from a belt-driven vane pump. To ensure adequate air from a belt-driven pump, more air, as a fraction of engine throughput, must be pumped at low speeds than at high speeds, Thus, the belt-driven pump must operate fast enough to supply enough air at low engine speeds. Then, if engine speed is increased, the belt-driven air pump supplies too much air and blowoff is required.

Exhaust Gas Recirculation. To control NOx to 1973 levels, exhaust gas recirculation (EGR) was included in the emission-control system. EGR has proven very effective to lower NOx by diluting the intake fuel and air mixture with a fraction of the exhaust gas. EGR lowers the peak temperature of the combustion event and slows the combustion rate. These effects combine to lower NOx, as shown in Fig. 12 for a 5.7 L V-8 engine tested on a test-stand dynamometer [Gumbleton, et al., 1974]. Percent of EGR is defined as the mass rate of recirculated exhaust gas divided by the sum of mass rates of air, fuel, and recirculated exhaust gas. With careful engine design, EGR rates greater than 20% are acceptable.

As shown in Fig. 12, measured BSNOx, brake specific NOx, g/kWh, fell drastically when EGR was increased to 15%. An EGR rate of 8% lowered NOx as much as a spark retard of 20°. As

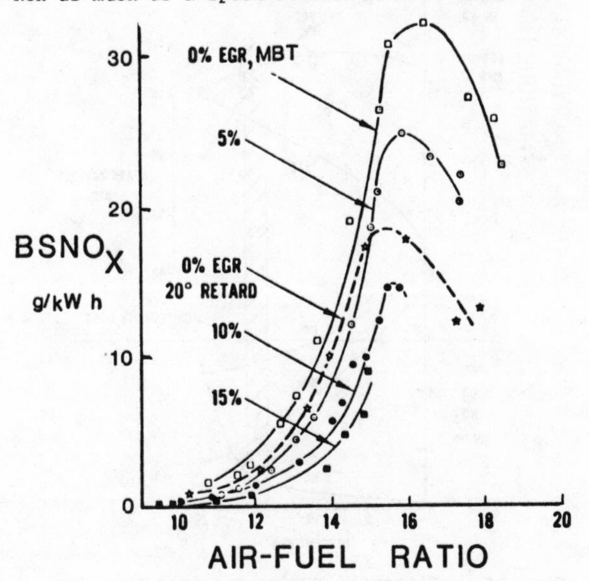

Fig. 12 - Effect of EGR on BSNOx for a 5.7 L V-8 Engine at Part Load

summarized in Fig. 13, the fuel economy improved slightly at rich A/F for EGR rates up to 15%, contrasting with the 10% fuel economy loss recorded when the spark timing was retarded 20°.

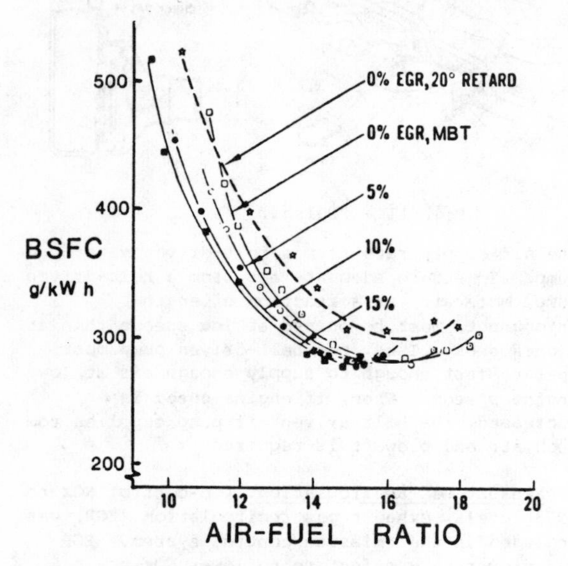

Fig. 13 - Effect of EGR on BSFC for a 5.7 L V-8 Engine at Part Load

Data for a 7.7 L V-8 operated at steady-state conditions of 88 km/h (55 mi/h) are plotted in Fig. 14, showing the effects of increased EGR on manifold vacuum, miles per gallon, spark timing, and output HC, CO, and NOx. Increasing EGR dilutes the fuel/air mixture, essentially lowering the maximum power output from an engine; thus the throttle opening must be larger to maintain output power. A larger throttle opening lowers manifold vacuum, as shown in Fig. 14.

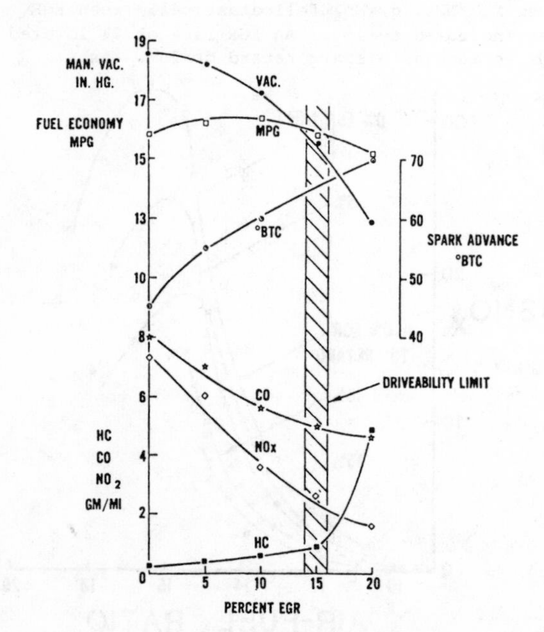

Fig. 14 - Effect of EGR on Engine Parameters for a 7.7 L V-8 Engine at Part Load with Best Economy Spark Advance

The dilution by EGR of the fuel/air charge also slows burn rate; thus spark timing must be advanced to obtain best-economy timing. For the engine condition tested, spark timing was advanced from 45° with no EGR to 65° with 15% EGR. With spark timing adjusted for best economy, fuel economy increased slighly in this engine as EGR increased up to a maximum at about 10% EGR. Emissions measured at the output from the engine show NOx decreased sharply with increased EGR, CO decreased somewhat, and HC increased slowly, until EGR levels were increased to about 15%. At 15% EGR, engine combustion deteriorated, causing the HC to increase sharply as the driveability of the vehicle deteriorated.

The amount of EGR that an engine can tolerate depends primarily on the rate of burning in the combustion chamber. A fast-burn chamber can tolerate more EGR that a slow-burn chamber. Much engineering effort has been expended to design combustion chambers to tolerate large amounts of EGR for control of NOx within the engine [Mattavi, 1980].

Control of EGR. Control of EGR and control of spark timing are interdependent for the best combination of fuel economy, emissions, and driveabiliy. Since high temperatures during combustion produce large quantities of NOx, the output of NOx increases with engine load [Gast, 1975]. Additionally, as load and speed are increased, a lower residual-gas fraction is trapped in the cylinder from the previous combustion event [Amann, 1980]. Effectively, this trapped gas is internal EGR, so the fraction of internal EGR is lowered as engine speed and load are increased. Thus to maintain total EGR rates, external EGR must be controlled, in part, as a function of engine speed and load.

A valve to control EGR can be mounted on the intake manifold to connect two passages, one passage in the intake manifold and the other passage leading to the exhaust manifold. Opening this EGR valve allows exhaust gas to flow into the intake manifold. Manifold vacuum acting on a diaphragm against a spring controls the position of the valve stem, thus controlling the flow rate of exhaust gas, as shown in Fig. 15 [Oldsmobile, Division, GMR, 1980].

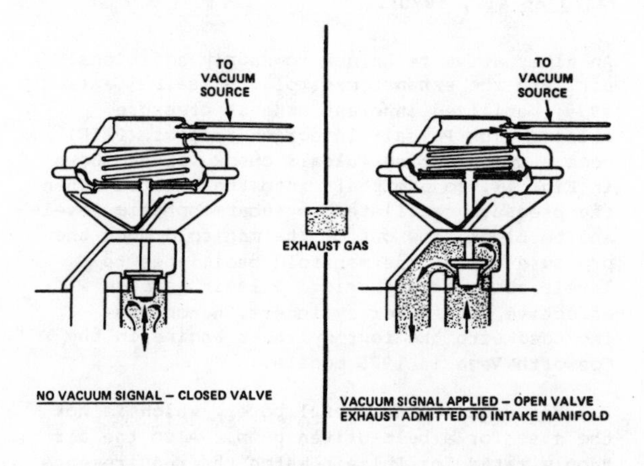

Fig. 15 - Vacuum Modulated EGR Valve

Engine backpressure increases with increased engine speed and load, so backpressure has been used to modulate EGR, i.e., more EGR is associated witn increased backpressure. Introduced in later years, a backpressure-modulated EGR valve is shown in Fig. 16 [Oldsmobile Division, GMC, 1980]. In this EGR valve, exhaust pressure communicates through the valve stem to control blockage of a bleed orifice in the main diaphragm. Thus, high exhaust pressure increases EGR rate, and low exhaust pressure decreases EGR rate. This control of the valve modulates EGR to increase with increased load on the engine.

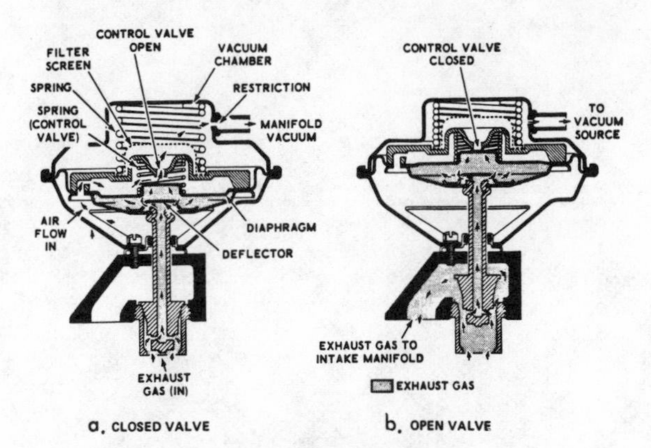

Fig. 16 - EGR Valve Modulated by Exhaust Backpressure

Closure for 1973 Emission Controls. By 1973, HC and CO were controlled to low levels by retarding spark timing. Retarded spark timing shortened engine warmup time, and also increased temperatures in the exhaust manifold to promote gas-phase reactions. To aid oxidation both during warmup and during fully warm operation, an air pump supplied additional air to the exhaust manifold. EGR emerged as a very effective technique to lower NOx. When spark advance was adjusted for MBT with EGR, fuel economy first increased, then decreased as EGR increased. The maximum level of EGR used was set short of the point of engine misfiring, called the lean limit.

SUMMARY

1. Sophisticated systems have evolved to control emissions from vehicles since automobiles began meeting exhaust emission standards in California in 1966 and nationwide in 1968.

2. Relative to the precontrol years of the mid 60's, tailpipe emisions from passenger cars through 1974 have been lowered 72% for unburned hydrocarbons, 67% for carbon monoxide and 24% for oxides of nitrogen. Crankcase, fuel tank, and carburetor evaporative losses have been lowered to negligible trace amounts.

3. To lower unburned hydrocarbons and carbon monoxide, early control systems quickened engine warmup and relied on post-engine oxidation reactions by using retarded spark timing and air injection.

4. Emission control by retarded spark timing lowers oxides of nitrogen substantially, lowers unburned hydrocarbons somewhat, and lowers carbon monoxide slightly at the expense of increased fuel consumption.

5. Exhaust gas recirculation can substantially lower oxides of nitrogen with minimal penalty in fuel economy.

REFERENCES

1. Amann, C. A., 1980, "Control of the Homogeneous-Charge Passenger-Car Engine -- Defining the Problem," SAE Paper 801440, SAE SP-477.

2. Gast, R. A., 1975, "Pulsair -- A Method for Exhaust-System Induction of Secondary Air for Emission Control, " SAE Paper 750172.

3. General Motors Corporation, 1971, Progress and Programs In Automotive Emissions Control.

4. General Motors Corporation, 1972, General Motors Emissions Control Systems Maintenance Manual.

5. General Motors Corporation, 1973, Report on Progress in Areas of Public Concern.

6. General Motors Corporation, 1976, General Motors Public Interest Report.

7. General Motors Corporation, 1981, General Motors Public Interest Report.

8. Gumbleton, J. J.; Bolton, R. A.; Lang, H. W.; 1974, "Optimizing Engine Parameters with Exhaust Gas Recirculation," SAE Paper 740104.

9. Herrin, R. J., 1978, "Emissions Performance of Lean Thermal Reactors -- Effects of Volume, Configuration, and Heat Loss," SAE Trans., Vol. 78, pp. 31-51 (Paper 780008).

10. King, J. B.; Schneider, H. R.; and Tooker, R. S.; 1970, "The 1970 General Motors Emission Control Systems, " SAE Paper 700149.

11. Mattavi, J. N., 1980, "The Attributes of Fast Burning Rates in Engines, " SAE Trans., Vol. 89, pp. 2783-2801 (Paper 800920).

12. Obert, E. F., 1950, Internal Combustion Engines Analysis and Practice, 2nd Ed., International Textbook Co., Scranton, PA.

13. Oldsmobile Division, General Motors Corporation, 1980, 1980 Oldsmobile Omega Service Manual.

14. Pao, H. C., 1982, "The Measurement of Fuel Evaporation in the Induction System During Warm-Up," SAE Paper 820409.

AN HISTORICAL OVERVIEW OF EMISSION-CONTROL TECHNIQUES FOR SPARK-IGNITION ENGINES: PART B – USING CATALYTIC CONVERTERS

J. R. Mondt
AC Rochester Division
General Motors Corporation
Flint, Michigan

ABSTRACT

Starting in the 1960's, spark-ignition engines have changed markedly in response to requirements for lowered exhaust and evaporative emissions. Specific emission requirements are set forth in Federal Regulations. Congress passed the Clean Air Act in 1963, and in 1965 added an amendment to regulate emissions from automobiles. Thus, automotive engineers embarked into an era of increased emphasis on engineering for control of emissions. In addition to changes made to basic engine designs, such as lowered compression ratios, reduced valve overlaps, and modified combustion chambers, substantial changes have been made in engine controls.

In this paper, developments after using catalytic converters are described. To meet stringent emissions requirements legislated for 1975, catalytic converters were introduced on all GM passenger cars in 1975 models. After that, on-board computers were introduced to control fuel metering, spark timing, idle speed, EGR, and on some vehicles, transmission shifts. On board computers ushered in an order of magnitude sophistication in engine control systems which have included, for example, TWC (three-way catalytic converter), TBI (throttle body injection), PFI (port fuel injection), SFI (sequential fuel injection) and mass airflow sensors.

INTRODUCTION

Congress passed the Clean Air Act of 1963, and two years later added provisions for control of automobile emissions. Adding legislated demands for emission control to the marketplace demands for fuel economy and driveability challenged the automotive industry as never before. Response to this challenge has been control of the engine by devices which limit emissions from the engine. Improving old devices and adding new devices required new technology and new terminology, all synthesized as the "emission-control system."

Part A of this two-part paper summarizes developments through 1974, prior to introducing catalytic converters. This paper, Part B, documents developments in emission control utilizing catalytic converters, both oxidizing and three-way. With three-way catalysts, onboard computers were introduced which added a new dimension to the control system for an automotive power plant. GM expenditures to meet emission standards for vehicles averaged 440 million for each of five years, 1976 through 1980.

EMISSION CONTROL SYSTEMS

As the standards for emissions became more stringent, emission controls became more sophisticated. Thus, a chronological structure for this paper seems logical, so the use of emission-control systems has been subdivided into three historical eras:

1. Precatalytic converter, prior to 1975
2. Oxidizing catalytic converter, 1975 through 1980
3. Three-way catalytic converter, 1981 through 1989

This paper summarizes oxidizing and three-way catalytic converter systems, after 1975.

<u>Oxidizing Catalytic Converter Era</u>

As shown in Table I, 1975 emission standards for NOx did not change compared with 1973; however, both HC and CO standards dropped about 50%. Spark control, improved fuel metering, faster engine warmup, and AIR were not sufficient to control HC and CO to the 1975 levels. Thus, oxidizing catalytic converters appeared on all GM cars in 1975.

TABLE I

Federal Exhaust Emission Requirements
1978 Test Procedure
(grams/mile)

	(1960)*	1970	1973	1975	1981-89
HC	(10.6)*	4.1	3.0	1.5	0.41 (0.41)**
CO	(84)*	34.0	28.0	15.0	3.4 (7.0)**
NOx	(4.1)*	---	3.1	3.1	1.0 (0.7)**

*uncontrolled
**California

1975 Emissions Controls: A GM emission-control system for 1975, shown schematically in Fig. 1, included the following additional items:

1. Oxidizing catalytic converter, OC
2. Improved early fuel evaporation system, EFE (also quick heat manifold, see Fig. 1)
3. High-energy ignition, HEI

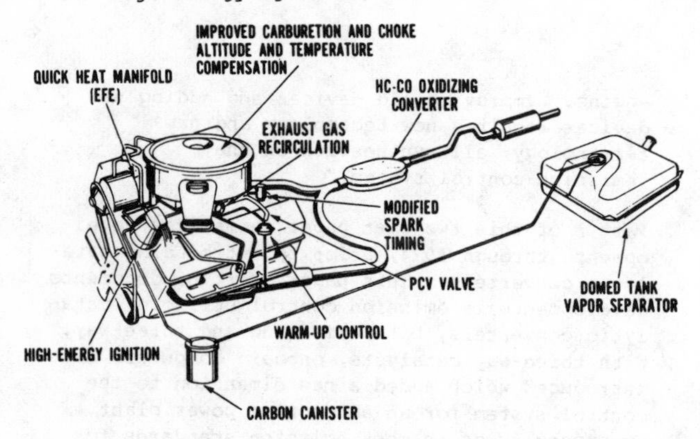

Fig. 1 - 1975 Emission-Control System

Oxidizing Catalytic Converter. Fundamentally, the oxidizing catalyst promotes the oxidation of HC and CO at much lower threshold temperature of approximately 270 C and 220 C respectively, as shown in Fig. 2, instead of the 750 C temperature needed for gas-phase oxidation. When a fresh catalyst is fully warm, conversion efficiency for CO attains 98 to 99%, and for HC attains 95%. Fully warm conversion efficiency is higher for CO than for HC because HC is a mixture of hydrocarbons containing a fraction of unreacting species, such as methane. Prior to adding the catalytic converter, emissions control systems relied on retarded spark timing to accomplish both lowered HC from the engine combustion, discussed previously, and higher temperatures in the exhaust manifold to promote oxidation of both HC and CO by gas-phase reactions. Unfortunately, retarding spark timing to lower emissions incurred a substantial penalty in fuel economy. Installing a catalytic converter to process the exhaust from a passenger-car engine represents an unparalleled step in the evolution of emission-control systems. Before the production converter system was finalized, hundreds of catalysts were tested for durability and

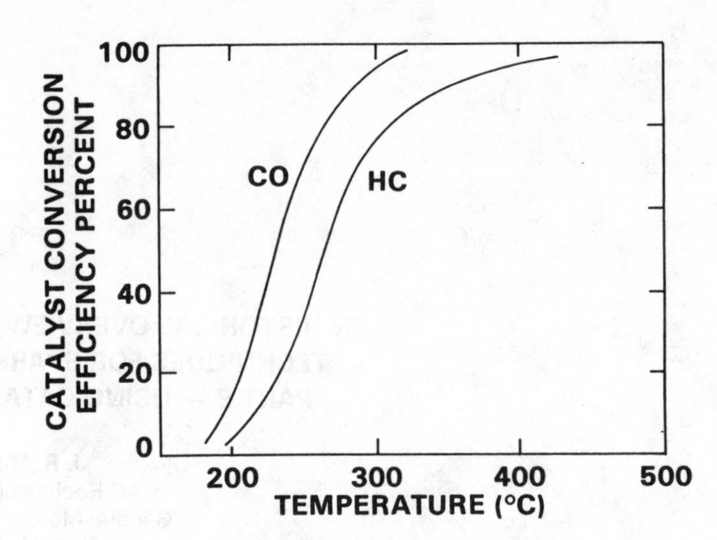

Fig. 2 - Lightoff for a Typical Oxidizing Catalyst

conversion efficiency. Millions of miles of vehicle testing and several alternative systems were evaluated before beaded catalysts were installed on all GM passenger cars in 1975 models [General Motors Corporation, 1973].

With a catalytic converter installed to oxidize HC and CO, spark timing could be calibrated with settings approaching MBT, the spark setting for best economy and maximum torque. Thus, both driveability and fuel economy improved when catalytic converters were added to the emission control system. The average fuel economy for the GM passenger car fleet increased from 12.0 MPG in 1974 to 15.4 MPG in 1975, as shown in Fig. 3.

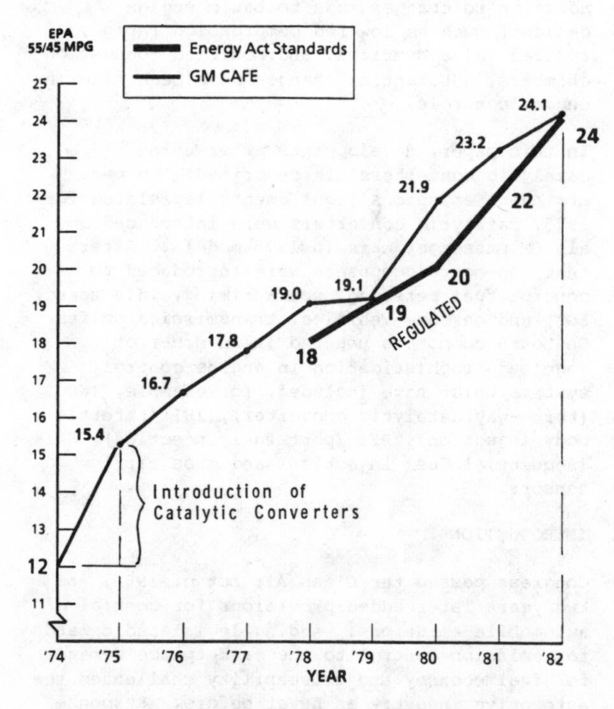

Fig. 3 - Sales Weighted Fuel Economy for GM Passenger Car Fleet

Platinum and palladium noble metals are used as catalysts to promote oxidation of HC and CO from automobile engines. These noble metals are very finely dispersed on a porous alumina coating on either a bead-bed or monolithic substrate.

For the bead-bed converter, shown schematically in Fig. 4, alumina is pelletized either as spheres or extrudates of approximately 3mm (0.12 in.) diameter. Platinum and palladium are impregnated into the pellets by submerging the pellets in a dilute solution of platinum and palladium salt. The pellets are retained in the converter by a perforated metal grid at both the inlet and outlet face of the converter.

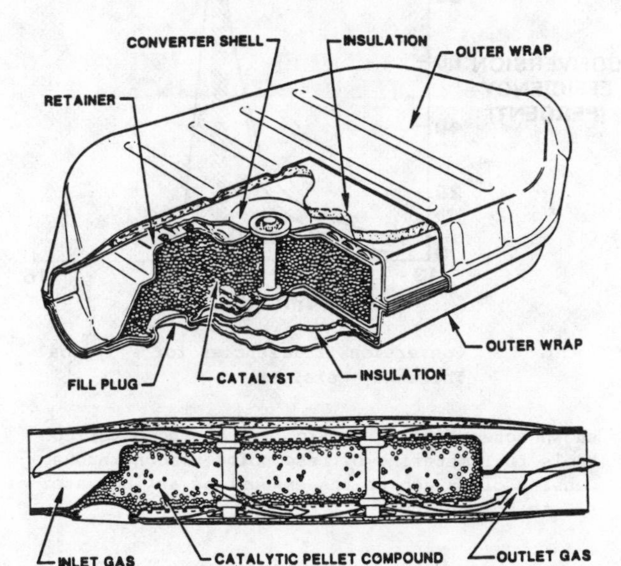

Fig. 4 - Pelletized Catalytic Converter

A typical monolith converter is made by extruding cordierite into a structure containing many parallel passages of square cross section, as shown in Fig. 5. The almost impervious cordierite walls are coated with a "washcoat" of porous alumina to provide support for the catalyst metals. A typical washcoat thickness varies from 15 to 25 microns. Monolith converters were phased into GM automobiles with small production volume in California in 1977, and increased production volume in 1980.

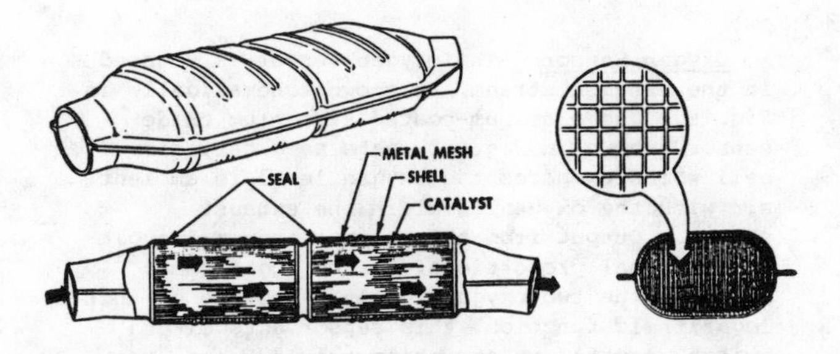

Fig. 5 - Monolith Catalytic Converter

Catalyst Aging. Both light-off and conversion performance of catalysts degrade with time in service. The amount of deterioration depends on the formulation of the catalyst, properties of the substrate, and properties of the exhaust gas supplied to the catalyst. Sintering and poisoning are two causes of degraded catalysts. Sintering agglomerates the catalyst metals and may cause chemical changes when temperatures exceed 850 C (1560 F) for a few hours. Poisoning can block active sites, and some poisons react chemically with the catalyst or substrate to degrade performance. Because of catalyst deterioration, catalytic converters are oversized to guarantee performance through an aging period on a vehicle of 80,000 km (50,000 mi).

Warmup Control. Quick warmup of a catalytic converter lowers emissions because essentially all emissions pass through the converter without reacting until the catalyst is heated to light-off temperature. Thus, to exploit the catalytic converter by quick warmup, more sophisticated systems using air injection, spark retard, and fast idle appeared on engines. Additional air from an air pump may be supplied during a cold start to ensure adequate oxygen for the oxidizing converter.

Early fuel evaporation (EFE) systems heat the carburetor and intake manifold to vaporize fuel droplets, and thus lower HC and CO during a cold start. Some EFE systems use a vacuum-operated valve to block flow in one section of the exhaust system, forcing hot exhaust gas through a passage in the intake manifold under the carburetor. This hot exhaust heats the carburetor and intake manifold and shortens the duration of cold-start enrichment. For some engines in later years, an electrical resistance heater has been used to evaporate directly liquid droplets of fuel entering the intake manifold [General Motors Corporation, 1981].

High-Energy Ignition. Fully warm catalytic converters oxidize over 95% of the HC supplied to them. This exothermic oxidation releases energy which can overheat the catalytic converter if the engine supplies excessive amounts of unburned HC, for example, as a result of engine misfiring. To minimize misfiring, or even partial misfiring, high-energy ignition (HEI) was employed with catalytic converter systems. To improve ignition with HEI, the voltage across the spark gap was increased to 35,000 volts from the previous 20,000 volts.

Closure for 1975 Emission Controls. Fig. 6 shows a vacuum hose schematic for a 1980 Oldsmobile Omega [General Motors Corporation, 1980], to summarize the sophisticated emission-control system for a vehicle equipped with a catalytic converter. With the carburetor as the focal point, other devices are shown schematically arranged as satellite components:

1. DS-VDV: distributor spark-vacuum delay valve
2. DS-VMV: distributor spark-vacuum modulator valve
3. EGR-TVS: exhaust gas recirculation-thermal vacuum switch
4. Pulsair unit for additional air
5. PCV: positive crankcase ventilation

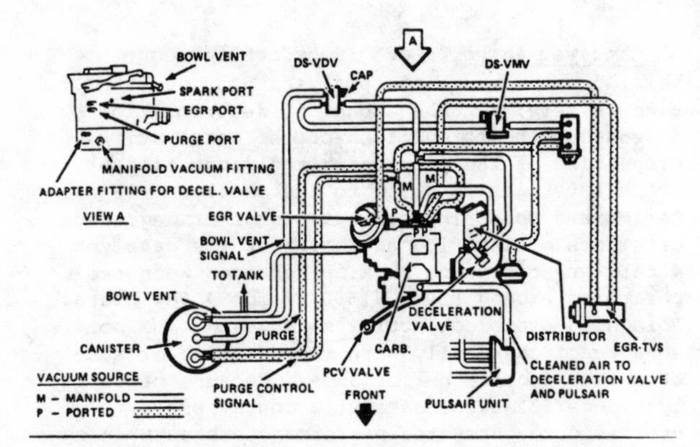

Fig. 6 - Vacuum-Hose, Schematic for a 1980 Oldsmobile Omega with 2.5 L L-4, Low Altitude, Automatic Transmission

6. Deceleration valve
7. Canister: charcoal canister for fuel evaporation control
8. EGR valve: exhaust gas recirculation valve

A side view of the carburetor body in the upper left of Fig. 6 shows several ports for vacuum, as well as the bowl vent which vents vapors to the charcoal canister. Spark port designates a vacuum source from a passage located part way up the side wall of the carburetor bore. This port is not exposed to manifold vacuum until the throttle plate opens far enough to pass the port. Thus, "ported vacuum" has no vacuum signal when the throttle is closed or just barely open.

Three-Way Catalytic Converter Era

As listed in Table I, allowable levels for HC, CO, and NOx were all lowered substantially in 1981. A cata- lyst approach for lowering all the pollutants in- volves oxidizing the HC and CO and simultaneously reducing the NOx. Several alternative systems were investigated before selecting a three-way catalyst (TWC) system. The three-way catalyst was so named because one catalytic converter can simultaneously oxidize and reduce to control all three exhaust pollutants: HC, CO, and NOx.

1981 Emission Controls: Components in the 1981 system, shown in Fig. 7 are:

1. Three-way catalytic converter (TWC)
2. Oxygen sensor
3. Electronic control module (ECM)
4. Electronic ignition
5. Closed-loop carburetor
6. Early fuel evaporation (EFE)
7. Exhaust gas recirculation (EGR)

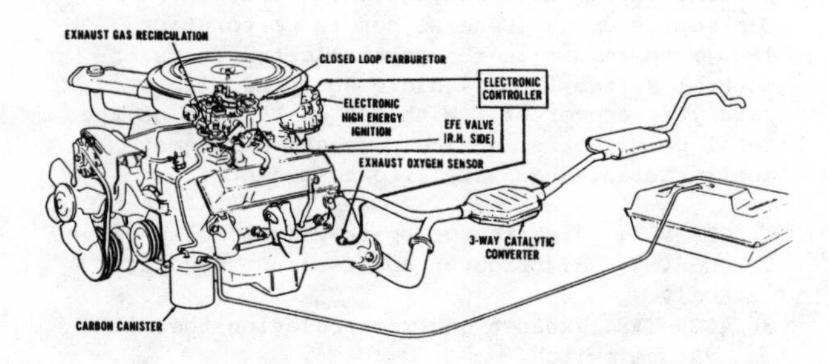

Fig. 7 - 1981 Emission-Control System

Three-Way Catalytic Converter. By adding rhodium to the platinum/palladium mixture, NOx can be reduced while simultaneously both HC and CO can be oxidized. Concurrent reducing and oxidizing is possible by controlling the air-fuel ratio within a narrow band of $\pm$ 0.2 A/F about stoichiometric A/F, as shown in Fig. 8. This narrow band in A/F is very difficult to achieve with a conventional carburetor. To achieve precise control of A/F, computer control of fuel metering was employed. This computer operates as part of a closed-loop system,

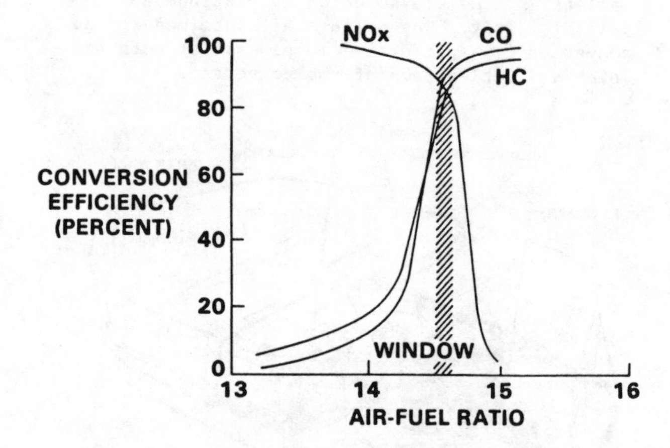

Fig. 8 - Conversion Efficiencies for a Typical Three-Way Catalyst

shown schematically in Fig. 9. The computer controls fuel metering by responding to continuous sensing of oxygen in the exhaust as an indicator of A/F in the engine.

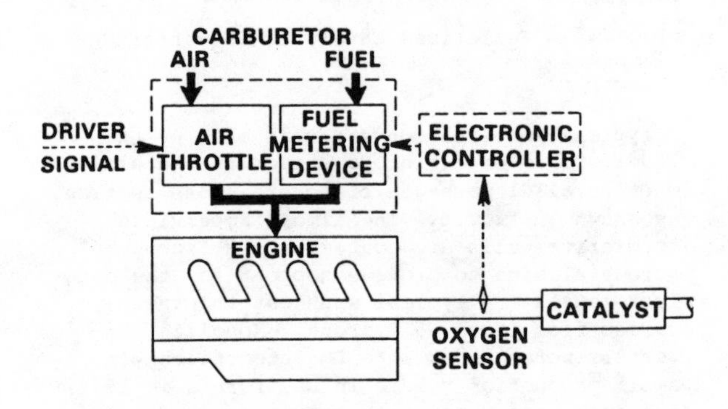

Fig. 9 - Schematic Diagram of a Closed Loop Control System

Oxygen Sensor. The oxygen sensor is mounted in the exhaust stream, as shown schematically in Fig. 9. The platinum-coated zirconium oxide sensor, shown in Fig. 10, acts as a galvanic cell which compares the oxygen level in ambient air with the oxygen level in the exhaust stream. Output from this sensor is a millivolt-level signal proportional to the logarithmic ratio of the two oxygen levels. Because of this logarithmic function, this sensor acts like a switch actuated at stoichiometric A/F and producing approximately 1000 mv when the exhaust is rich and 10 mv when the exhaust is lean.

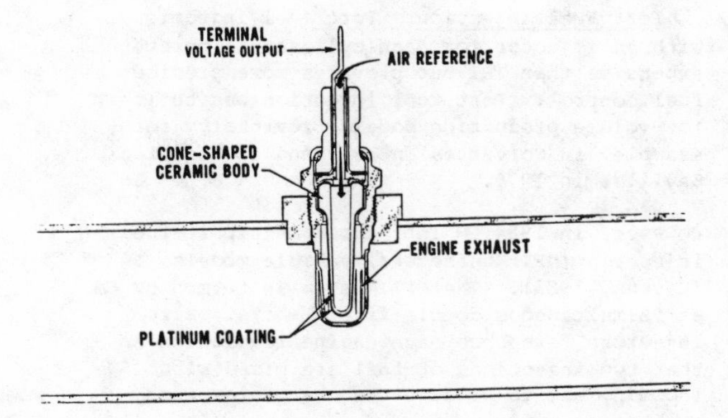

Fig. 10 - Oxygen Sensor in the Exhaust Pipe

Electromechanical Carburetor. To meter the fuel for this closed-loop system, an electromechanical carburetor was employed by GM in California is a few 1979 and 1980 models, and nationwide in 1981. The electromechanical carburetor controls fuel input by oscillating two metering rods at a constant frequency of 10 Hertz. A main and an idle metering rod, shown in the carburetor cross section, Fig. 11 [Hegedus and Gumbleton, 1980], oscillate between a lean and a rich stop. The length of time spent at either stop can be varied from 10 to 90%, thereby providing pulse-width modulated control of A/F.

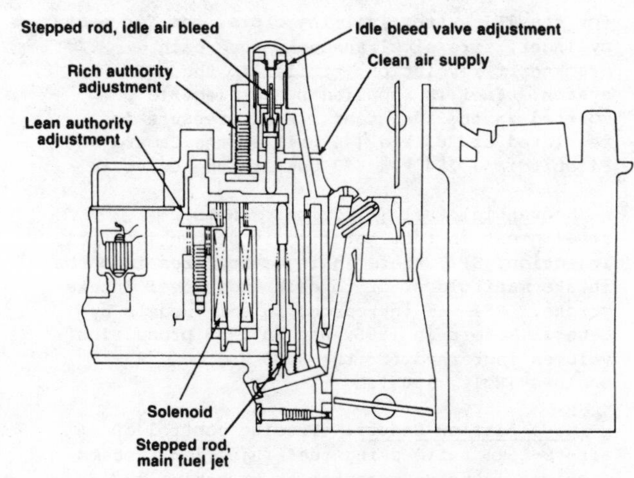

Fig. 11 - Electromechanical Carburetor for Closed-Loop Emission Control

Oscillation of air-fuel ratio can widen the A/F range for conversion of all three species because of oxygen "storage" by the catalyst during a lean period. "Stored" oxygen can be subsequently released to oxidize HC and CO even during a period of rich A/F. Engineers continue studies of A/F dynamics to exploit this "storage" phenomenon for high conversion efficiencies.

Dual-Bed Catalytic Converter. Development has indicated that for a given conversion, a smaller total volume of catalyst is required for a dual-bed configuration, as shown in Fig. 12, then for a single-bed converter. The front bed contains a three-way catalyst, and the rear bed

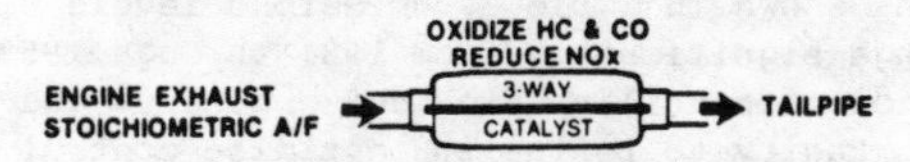

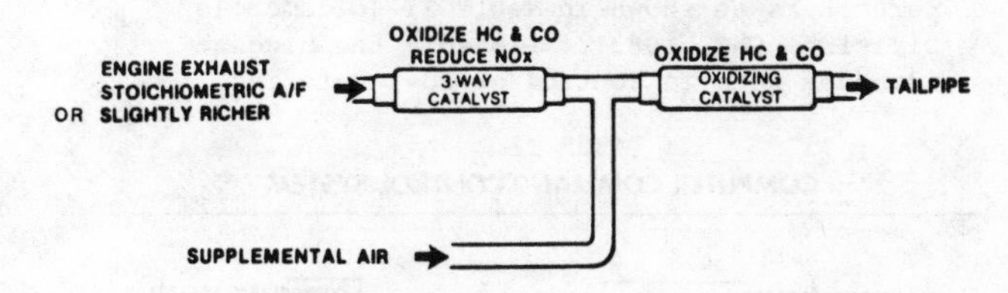

Fig. 12 - Three-Way and Dual Converter Systems

contains an oxidizing catalyst. Additional air from an air pump injected upstream of the rear bed assures oxidizing conditions in the rear bed.

Operation of oxidizing catalysts, three-way catalysts, and oxygen sensors involve noble-metal catalysts. Inherent in any catalyst is deterioration from poisoning, sintering, and plugging. Deterioration with aging compromises the performance of all of these catalytic systems.

Complex Systems. The realm of computer control expanded to include many parameters. An advanced system is shown schematically in Fig. 13 [Grimm, et al., 1980], where the computer controls the following:

1. Electromechanical carburetion
2. EST - electronic spark timing
3. Idle speed
4. Air management
5. EGR - exhaust gas recirculation
6. Torque converter lock up

Closure for 1981 Emission Controls. Thus, by 1981, the emissions standards had decreased to levels met by employing a three-way catalyst. Precise control of A/F for the three-way catalysts was achieved with Computer Command Control. This closed-loop system adjusts fuel metering to maintain precise control of A/F in the exhaust stream. Early computer systems controlled only fuel metering, spark timing, and idle speed.

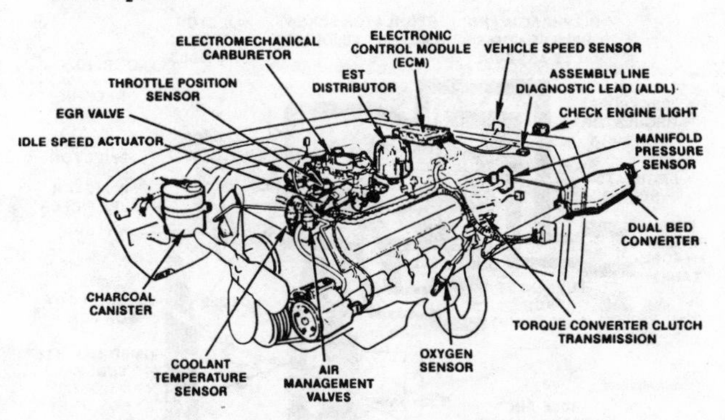

Fig. 13 - A Typical 1981 GM Computer Command Control System

Developments from 1982 through 1989

As shown in Table I, emissions levels did not change significantly from 1981 through 1989. A period of stability provided engineers the opportunity to refine and optimize control systems to control emissions and also provide improved fuel economy and driveability. Use of on-board computers expanded to control many parameters as shown in Table II [Oldsmobile Division, GMC, 1983]. Probably the biggest advances were in control of fuel metering.

TABLE II
COMPUTER COMMAND CONTROL SYSTEM

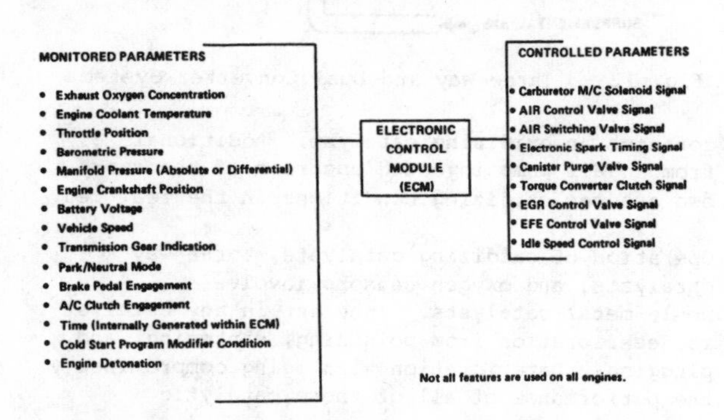

More precise fuel metering simultaneously lowers emissions, increases fuel economy and improves driveability. To significantly improve fuel metering, fuel-injection systems have been developed and a mass airflow sensor has been developed.

Throttle Body Injection. To improve accuracy of control of fuel metering, throttle body injection, Fig. 14, was introduced in a Cadillac in 1980, with additional production in other models in 1981 and 1982 [McElray, 1981]. The electronic control module, ECM, responds to inputs from several sensors and controls pulse width of the solenoid-operated injector to spray the correct amount of fuel into the throttle body just upstream of the throttle plate. Small engines had one injector and large engines had two injectors. An electric fuel pump mounted in the fuel tank delivers more gasoline than is needed. The excess gasoline circulates through the injectors and pressure regulator cooling them before returning to the tank. A pressure regulator controls fuel pressure to 70 kPa at the injector.

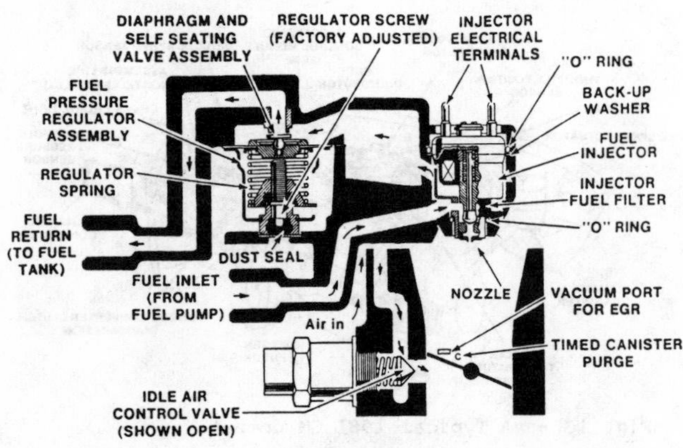

Fig. 14 - Throttle Body Injection

Port Fuel Injection. Port fuel injection with an injector for each cylinder is more expensive than TBI but provides more precise fuel control. Port fuel injection was built in low-volume production models previously, for example, in Corvettes in 1957 and in Cadillac Sevilles in 1975.

However, in 1984 GM introduced multiport fuel injection (MFI) on several vehicle models [Givens, 1983]. The MFI system is termed by GM as "simultaneous double-fire" -- i.e., all injectors fire once each engine revolution so that two injections of fuel are mixed with incoming air to produce the charge for each combustion cycle.

Each injector is solenoid operated from electric pulses supplied by the ECM. The ECM controls fuel injection after monitoring inputs from an exhaust-gas oxygen sensor, coolant temperature sensor, detonation sensor, throttle position sensor, and in some systems, the hot-film mass airflow sensor.

In 1985, tuned port injection (TPI) fuel induction was introduced on two GM engines -- the 5.0 L V-8 Camaro and Firebird, and the 5.7 L V-8 Corvette [Givens, 1984]. The TPI system results in significant increases in both torque and power by tuning inlet air runners to take advantage of air pulses caused by opening and closing the intake valves.

For the TPI system, all injectors, one for each cylinder, fire simultaneously once each crankshaft revolution, similar to the MFI system. Fuel is supplied by an electric pump mounted in the fuel tank. Fuel pressure is regulated at 303 kPa (44 psi) in the Camaro and Firebird and 254 kPa (37 psi) in the Corvette.

Sequential Fuel Injection. Another development in fuel metering is sequential fuel injection, SFI, where an injector fires into the intake manifold prior to each individual intake stroke. SFI was introduced in some models by General Motors in 1985, and in 1988 production volumes increased for the 3800 and Quad 4 engines [Holt, 1987].

Mass Airflow Sensor. Precise control of air-to-fuel ratio using fuel-injection systems requires accurate measurement of engine airflow. Airflow to the engine can be measured by an airflow sensor, sometimes located between the air filter and the throttle body [Givens, 1983]. A heated thin flat plate senses mass airflow by determining the amount of energy required to maintain the flat plate temperature 42 C (75 F) above the temperature of the incoming air. A bypass channel is provided to supply the amount of air needed for idle.

Electronically Controlled EGR. A system to control EGR electronically was introduced on some 1985 models using the Exhaust Vacuum Regulator Valve (EVRV) shown in Figure 15 [Givens, 1984]. Opening of the valve is controlled by the vehicle ECM. After monitoring r/min, load, torque, converter-clutch engagement, and engine temperature, appropriate variables are sent from the ECM to the

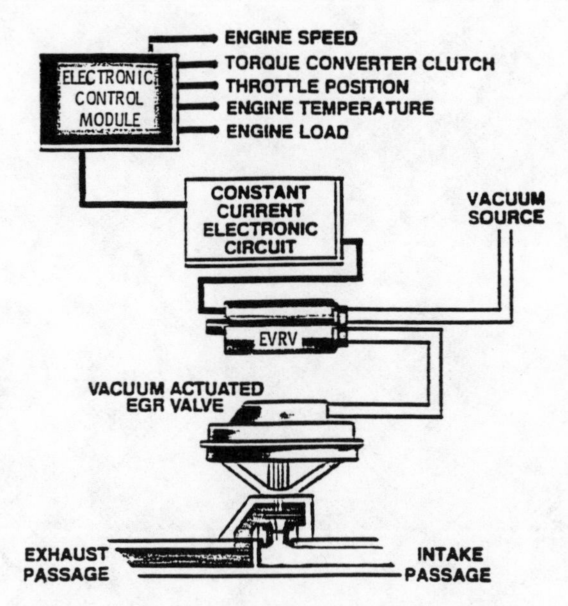

Fig. 15 - Electronically Controlled EGR

constant-current electronic circuit (CCEC). The CCEC interprets ECM data and transmits signals to the EVRV. The EVRV is turn modulates vacuum communicated to the EGR valve to continuously adjust the valve to control EGR rate.

Modern Engine Technology. The Quad 4 engine included modern product and process technologies with four valves per cylinder, overhead cams, and tuned intake and exhaust system [Thompson, et al., 1987]. With four valves, valve timing can be controlled to improve combustion and also control cylinder charge residuals. With this feature, the external EGR valve was eliminated and internal EGR along with a three-way catalyst control NOx emissions.

Closure for Emission Controls, 1982-1989. With emissions levels stabilized, developments were focused to improve performance and lower fuel consumption. Use of on-board computers expanded, especially for control of fuel metering. Computer control poses new opportunities for automotive engineers, especially those working on control of emissions.

SUMMARY

1. To achieve 1975 standards, oxidizing catalytic converters were employed, allowing improved spark advance calibration, which contributed significanttly to the increase in weighted-average fuel economy from 12.0 to 15.4 miles per gallon realized by GM that year.

2. Starting with 1981 models, three-way catalytic converters were employed by GM to reduce oxides of nitrogen while simultaneously oxidizing both unburned hydrocarbons and carbon monoxide.

3. Precise control of A/F for the three-way catalyst was achieved by computer control of fuel metering in a closed-loop system responding to the output from an oxygen sensor in the exhaust.

4. Since catalytic converters and the oxygen sensor both depend on catalysts, which degrade with aging in service, the emission control system must be designed to tolerate deterioration and continue to function satisfactorily for the mandated life of 80,000 km.

5. Using on-board computers, fuel metering developments included TBI, MFI, SFI, and TPI fuel injection systems to simultaneously control emissions while improving driveability and lowering fuel consumption.

6. In modern passenger cars, the computer controls most of the engine operation, all of the emission-control components, and certain vehicle components, such as transmission shifts.

7. New engine designs include control of emissions as a primary consideration.

REFERENCES

1. General Motors Corporation, 1973, Report on Progress in Areas of Public Concern.

2. General Motors Corporation, 1980, Oldsmobile Omega Service Manual.

3. General Motors Corporation, 1981, Chevette Shop Manual.

4. Givens, L., 1983, "Technical Highlights of the 1984 Automobiles," SAE Automotive Engineering, Vol. 91, No. 10, pp. 39-57.

5. Givens, L., 1984, "Technical Highlights of the 1985 Automobiles, " SAE Automotive Engineering, Vol. 92, No. 10, pp. 39-51.

6. Grimm, R.A., Bremer, R.J., and Stonestreet, S.D., 1980, "GM Micro-Computer Engine Control System, " SAE Paper 800053.

7. Hegedus, L.L., Gumbleton, J.J., 1980, "Catalysts, Computers, and Cars: A Growing Symbiosis, " Ceramics Engineering and Science Proceedings of the 9th Materials Conference, American Chemical Society, pp. 403-428.

8. Holt, D.J., 1987, "Technical Highlights of the 1988 Automobiles, " SAE Automotive Engineering, Vol. 95, No. 10, pp. 44-71.

9. McElroy, J., 1981, "New Bonneville and TBI at Pontiac," Automotive Industries, Vol. 161, No. 10, p. 69.

10. Oldsmobile Division, General Motors Corporation, 1983, 1984 Oldsmobile New Product Service Information Manual.

11. Thompson, M.W., Frelund, A.R., Pallas, M., and Miller, K.D., 1987, ""General Motors Quad 4 Engine," SAE Paper 870353.